BIOCHEMISTRY RESEARCH TRENDS

AF734418

ISOPROSTANES AND NEUROPROSTANES

PRODUCTION AND ROLE IN DIFFERENT PATHOLOGIES

BIOCHEMISTRY RESEARCH TRENDS

Additional books in this series can be found on Nova's website under the Series tab.

Additional e-books in this series can be found on Nova's website under eBooks tab.

BIOCHEMISTRY RESEARCH TRENDS

ISOPROSTANES AND NEUROPROSTANES

PRODUCTION AND ROLE IN DIFFERENT PATHOLOGIES

JAVIER MARHUENDA HERNÁNDEZ, PHD
DÉBORA VILLAÑO VALENCIA, PHD
BEGOÑA CERDÁ, PHD
AND
PILAR ZAFRILLA RENTERO, PHD

NOVINKA

New York

Copyright © 2017 by Nova Science Publishers, Inc.

All rights reserved. No part of this book may be reproduced, stored in a retrieval system or transmitted in any form or by any means: electronic, electrostatic, magnetic, tape, mechanical photocopying, recording or otherwise without the written permission of the Publisher.

We have partnered with Copyright Clearance Center to make it easy for you to obtain permissions to reuse content from this publication. Simply navigate to this publication's page on Nova's website and locate the "Get Permission" button below the title description. This button is linked directly to the title's permission page on copyright.com. Alternatively, you can visit copyright.com and search by title, ISBN, or ISSN.

For further questions about using the service on copyright.com, please contact:
Copyright Clearance Center
Phone: +1-(978) 750-8400 Fax: +1-(978) 750-4470 E-mail: info@copyright.com.

NOTICE TO THE READER

The Publisher has taken reasonable care in the preparation of this book, but makes no expressed or implied warranty of any kind and assumes no responsibility for any errors or omissions. No liability is assumed for incidental or consequential damages in connection with or arising out of information contained in this book. The Publisher shall not be liable for any special, consequential, or exemplary damages resulting, in whole or in part, from the readers' use of, or reliance upon, this material. Any parts of this book based on government reports are so indicated and copyright is claimed for those parts to the extent applicable to compilations of such works.

Independent verification should be sought for any data, advice or recommendations contained in this book. In addition, no responsibility is assumed by the publisher for any injury and/or damage to persons or property arising from any methods, products, instructions, ideas or otherwise contained in this publication.

This publication is designed to provide accurate and authoritative information with regard to the subject matter covered herein. It is sold with the clear understanding that the Publisher is not engaged in rendering legal or any other professional services. If legal or any other expert assistance is required, the services of a competent person should be sought. FROM A DECLARATION OF PARTICIPANTS JOINTLY ADOPTED BY A COMMITTEE OF THE AMERICAN BAR ASSOCIATION AND A COMMITTEE OF PUBLISHERS.

Additional color graphics may be available in the e-book version of this book.

Library of Congress Cataloging-in-Publication Data

Names: Marhuenda Hernbandez, Javier. | Villadno Valencia, Dbebora. | Cerdba, Begodna. | Zafrilla Rentero, Pilar.
Title: Isoprostanes and neuroprostanes: production and role in different pathologies /
editors, Javier Marhuenda Hernbandez, Dbebora Villadno Valencia, Begodna Cerdba,
and Pilar Zafrilla Rentero (Catholic University of San Antonio, Murcia, Spain).
Description: Hauppauge, New York: Nova Science Publisher's, Inc., [2016] |
Series: Biochemistry research trends | Includes bibliographical references and index.
Identifiers: LCCN 2016033955 (print) | LCCN 2016046302 (ebook) | ISBN 9781634852531 (softcover) |
ISBN 9781634859424 (ebook) | ISBN 9781634859424
Subjects: LCSH: Oxidation, Physiological. | Isoprostanes.
Classification: LCC QP177 .I86 2016 (print) | LCC QP177 (ebook) | DDC 572/.53--dc23
LC record available at https://lccn.loc.gov/2016033955

Published by Nova Science Publishers, Inc. † New York

CONTENTS

PREFACE

Eicosanoids are a group of molecules originating from lipid character oxidation of essential fatty acids of 20 carbon atoms. All eicosanoids are molecules of 20 carbon atoms and are grouped into prostaglandins, thromboxanes, leukotrienes, hydroxy acids and certain leukotriene precursors. Are molecules involved in complex networks of cellular communication animal organism, including humans? Other products can also be formed by the oxidation of different polyunsaturated fatty acids (PUFAs).

Reactive oxygen species (ROS) and/or nitrogen reactive species (NOS) lead to oxidative reactions, causing oxidative stress in the organism. Under such conditions, many molecules and structures are susceptible to oxidation, causing dysfunctions and certain disorders in the organism. Lipids, especially PUFAs, are easily oxidized (lipid peroxidation).

Lipid peroxidation products have been investigated in order to determine their use as biomarkers of oxidative status. Linolenic acid (ALA), arachidonic acid (AA), adrenic acid (AdA), eicosapentaenoic acid (EPA) and docosahexaenoic acid (DHA) are commonly studied fatty acids that result in Phytoprostanes (PhytoPs). Isoprostanes (IsoPs), F_2-dihomo-isoprostanes (F_2-dihomo-IsoPs), F_3-Neuroprostanes (F_3-NeuroPs) and F_4-Neuroprostanes (F_4-NeuroPs), respectively.

IsoPs have been established as general and reliable "in vivo" oxidative stress biomarker, because of their link to oxidative stress, and their presence in several clinical physiological and pathological conditions such as aging, cardiovascular disorders, cancer, and neurological pathologies. They are prostaglandin-like compounds, produced by non-enzymatic peroxidation of AA and catalyzed by free radicals or by esterification of cell membrane lipids. Moreover, central nervous system structures like neurons are also oxidizable due to their abundant content of PUFAs. An increase in ROS/NOS could cause neural membrane damage, leading to the development of different neuropathologies - including Alzheimer's disease (AD), Parkinson's disease (PD), or multiple sclerosis (MS).

Arachidonic acid is widely distirbuted along the whole organism, while adrenic acid, eicosapentaenoic acid and docosahexanoic acid are mainly found in the central nervious system. Therefore, while the mesurement of F_2-isoprostanes allows knowing the oxidative damage produced at sistemic level, neuroprostanes and F_2-dihomo-isoprostanes are representative of oxidative atack to the central nervious system.

Lipid peroxidation products have been investigated in order to determine their use as biomarkers of oxidative status in the human body. F_2-dihomo-IsoPs have been related to early brain damage and white matter injury, and Neuroprostanes affords the dimension of neuronal oxidative damage. Because of their link to oxidative stress, and their presence in several clinical entities such as aging, cardiovascular disorders, cancer, and neurological pathologies, Isoprostanes have been established as general and reliable in vivo oxidative stress biomarker. Prostaglandins and Isoprostanes are considered as lipid mediators, involved in the physiopathology of all organs. Therefore, the comprehension about the formation of these molecules from oxidative attack could be an effective weapon for the maintenance and development of human health.

The present book aims to resume the current knowledge about the formation rute of IsoPs, NeuroPs, F_2-dihomo-IsoPs and their subsecuent effects on the organism.

Dr. Javier Marhuenda Hernández, PhD

Chapter 1

OXIDATIVE STRESS AND DISEASES

Pilar Zafrilla Rentero, PhD

ABSTRACT

Oxidative stress has been associated with diverse pathophysiological events, including cancer, renal disease, and neurodegeneration. More recently, it has become apparent that reactive oxygen species (ROS) also plays a role in the development of vasculopathies, including those that define atherosclerosis, hypertension, and restenosis after angioplasty.

ROS regulates several general classes of genes, including adhesion molecules and chemotactic factors, antioxidant enzymes, and vasoactive substances. Some of these are clearly an adaptive response, such as the induction of superoxide dismutase and catalase by H_2O_2.

When the physiological redox state of cells is disturbed, ROS can severely affect cellular signalling and function. Indeed ROS, which are not neutralized or scavenged by antioxidant molecules, may react with nucleic acids and proteins and thereby alter the biochemical and physical properties of these important cellular components. Moreover, the exposure of lipids to free radicals induces a non-enzymatic reaction cascade resulting in an increased formation of bioactive molecules named isoprostanes. Consequently, systemic isoprostane formation is significantly increased in a variety of pathological processes associated with oxidative stress. Isoprostanes participate in oxidative injury by modulating platelet activation and adhesion and by reducing the antiplatelet activity of NO.

Keywords: oxidative stress, antioxidants, oxygen reactive species, disease, health promotion

Introduction

Oxygen radicals are involved in many biochemical activities of cells such as signal transduction, gene transcription and regulation of soluble guanylate cyclase activity. Nowadays, population is repetitively exposed to free radicals from the environment such as pollutants and cigarette smoke. Natural resources such as radon, cosmic radiation, as well as cellular metabolisms (respiratory burst, enzyme reactions) are also able to release free radicals into the environment.

As explained by Fang et al. (2002), "Overproduction of free radicals can cause oxidative damage to biomolecules, (lipids, proteins, DNA), eventually leading to many chronic diseases such as atherosclerosis, cancer, diabeties, cardiovascular diseases, chronic inflammation, strokes, aging and other degenerative diseases in humans". Neurodegenerative diseases, such as Alzheimer's disease (AD), Parkinson's disease (PD) and Amyotrophic lateral sclerosis (ALS), are defined by the progressive loss of specific neuronal cell populations and are associated with protein aggregates. A common feature of these diseases is the oxidative damage of neurons, which might be responsible for the dysfunction or death of neuronal cells that contributes to disease pathogenesis.

Antioxidants can be divided in exogenous (natural or synthetic) and endogenous. Both are responsible for the elimination of free radicals or scavenging ROS and their precursors, stopping formation of ROS and binding metal ions required for catalysis of ROS generation.

Despite the intake of natural antioxidants it has demonstrated to reduce risk of cancer, cardiovascular diseases or diabetes, although there is great debate in this matter [2].

1.1. Oxidative Stress

Oxidative stress is caused by an imbalance between the production of free radicals and antioxidant defenses in favor of the oxidation and leads to lipid peroxidation, membrane protein, and DNA damage, which is thought to be important in the pathogenesis of a variety of neurological disorders, especially neurodegenerative diseases or atherosclerosis, cancer, and aging [3, 4].

There are two reactive oxygen species (ROS) subgroups; free radicals such as superoxide radicals (O_2^-) and non-radical ROS such as hydrogen peroxide (H_2O_2). Both radicals ROS and non-radical ROS are common in the presence of an oxygen atom, which differentiates them from the reactive nitrogen species (RNS). ROS can be found in the environment, such as in pollutants, tobacco smoke, iron salts, and radiation, or it can be generated inside cells, and this can happen through multiple mechanisms. Generally, mitochondria are the most important sources of cellular ROS where continuous production of ROS takes place. This is the result of the electron transport chain located in the mitochondrial membrane, which is essential for the energy production inside the cell. Under normal conditions, due to protection, the inner membrane of mitochondria incorporates numerous free radicals neutralizing molecules called scavengers and antioxidant enzyme systems [5, 6].

Lipid peroxidation is the most important source of free radical-mediated injury that directly damages neuronal membranes, yielding a number of secondary products responsible for extensive cellular damage. Any specific repair process of lipid peroxidation does not exist as it does for proteins and DNA. That fact may explain why moderate levels of lipid peroxidation could have physiological significance for cell signaling and membrane remodeling. It occurs due to an increased generation and/or reduced elimination of reactive species (RS) by the antioxidant defense system [7].

Free radicals are molecules with a paired electron in their outer orbit. They have very important role in the origin of life and biological evolution, leaving beneficial effects on the organisms. Oxygen radicals are involved in many biochemical activities of cells such as signal transduction, gene transcription and regulation of soluble guanylate cyclase activity [8].

ROS are short-lived species. Consequently, their measurement within integrated systems, such as animal models and humans, has proven to be a complex challenge. Traditionally, ex vivo indices, such as the oxidizability of LDL or spin trapping approaches, have been deployed, but increasingly, attention has focused on the development of in vivo biomarkers of oxidant stress. Essentially, the approach has been indirect and configured on the identification of chemically stable, free radical–catalyzed products of lipid peroxidation (Figure 1.1.) (such as isoprostanes), modified proteins (such as nitrated fibrinogen), and indices of free radical–catalyzed modification of DNA (such as 8-oxo-deoxyguanosine) [9, 10].

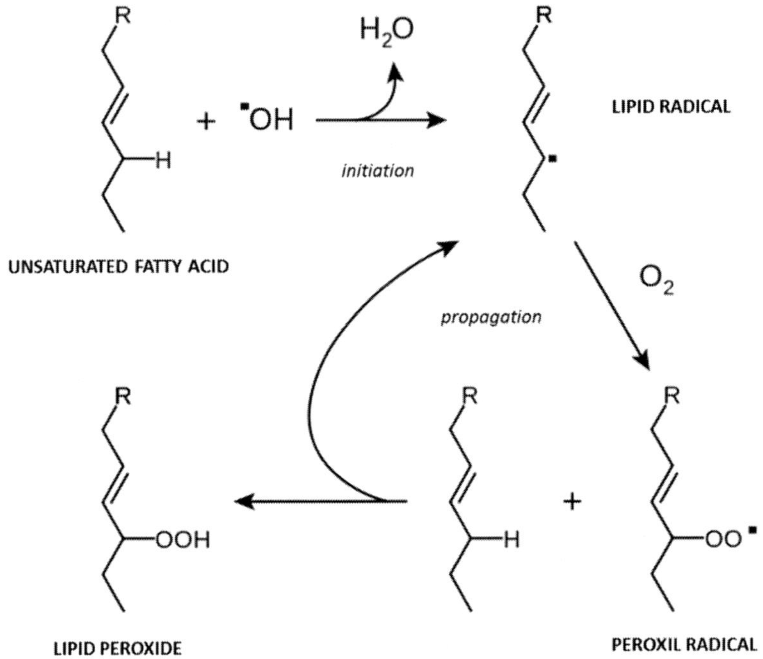

Figure 1.1. Peroxidation of fatty acids.

Oxidative stress is usually associated with oxidative damage, which is defined as "the biomolecular damage caused by an attack of ROS upon the constituents of living organisms" [11]. The ability of cells or tissues to withstand oxidative stress is largely dependent on the efficiency of the overall antioxidant defense system to scavenge excess RS, without compromising the physiological roles of ROS [3, 11].

1.2. Antioxidants

The terminus antioxidant refers to any compound capable to both delay or prevent the oxidation of a oxidizable substrate, in a minor concentration of such substrate [4, 11]. The functions of antioxidants involve reducing oxidative stress, mutations on DNA and RNA or malformations, among others. The scientific literature has proved the ability of antioxidants to reduce the effects of ROS injury, diminishing the development of cancer and degenerative diseases. Nonetheless, the sustained free radical injury can overcome the antioxidant capacity of the organism leading to the occurrence of different diseases [4]. The effects exerted by antioxidant is comprised of both preventing ROS event and scavenging the radicals that are formed [12]. The enzymatic procedure consists of repairing damaged DNA and RNA, eliminating oxidized proteins, and repairing oxidized lipids [13]. The decline of the repairing procedures leads to aging and degenative, due to an oxidative attack produced by ROS event [3].

The antioxidants can exert their effects at different stages of the oxidative radical process, which involves the sequential steps defined as initiation, propagation and chain termination [14]. The mechanism of the antioxidant for preventing oxidative attack comprises of: dimishing molecular oxygen, removing oxidant metalic ions, duping O_2^- and H_2O_2, breaking the chain of a radical arrangement or reducing the concentration of singlet oxygen (O_2^-) [13]. Antioxidants which inhibit lipid peroxidation by removing oxygen, by reducing the O2 concentration, or by breaking transitional metal ions, are named preventive antioxidants. Meanwhile, the compunds that are able to break the oxidant chain reactions, scavenging free radical species, are named primary antioxidants. Other antioxidants, named as secondary antioxidants, are: singlet oxygen deactivators, peroxide decomposers producing non-radical species, transition metal chelators, inhibitors of oxidative enzymes or UV emission absorbers [13].

It is important to note that secondary antioxidants can show synergetic properties combined with primary antioxidants, leading to many other antioxidant mechanisms [15]. The synergetic effects are stabilizing primary antioxidants by an acidic environment, chelating oxidative metals and restoring primary antioxidants by hydrogen donation from secondary antioxidants.

When the concentration of antioxidant is less than the pro-oxidant species, it could lead to the development of many diseases such as cancer, cardiovascular diesases, neurodegerative diseases or diabetes, among other [13].

1.3. Relation of Oxidative Stress and Diseases

1.3.1. Oxidative Stress and Cancer

Oxidative stress, as a disbalance between oxidant and antioxidants favouring the oxidants, involves the injury of proteins, DNA and lipids, resulting in cell death [16].

Cancerous cells contain significantly higher amounts of ROS than healthy cells [17]. Moreover, some oncogenes as Jun and Fos are stimulated by ROS, increasing the probability of developing malignant tumor. In fact, overexpression of Jun genes are associated to lung cancer [18]. Peroxidation products of polyunsaturated fatty acids are intermediate metabolites resulting in the accumulation within cell membranes. That accumulation of peroxides could lead to inflammation and fibrosis of strutures [19].

Additionally, changes in protein and lipid arrangements are related to an increased risk of mutagenesis, caused by lipid peroxidation. The prooxidant molecules are able to interact with DNA, modificating DNA polymerases or inhibiting the DNA repair system [20].

There are several studies that relate oxidative stess or an intake of antioxidant to the development and occurrence of some kind of cancer. Raafat et al. (2011) reported the importance of the intake of quinoline sulfonamide (a potent radioprotective antioxidant) and the incidence and development of cancer in mice. They found that the injuring effects and the occurance of cancer caused by exposure to g-radiation could be controlled by an administration of the antioxidant prior to irradiation.

Other studies measured the effect of carbon tetrachloride (CCl_4) free radicals. The metabolites formed from CCl_4 as chloromethyl ($\cdot CCl_3$) and trichloromethyl peroxyl ($OOCCl_3$) both exerted potent hepatotoxin effects, inducing morphological changes in many of the cell structures [21]. Moreover, the induction of tetrachloride cirrhosis in rats leads to an increase in plasmatic levels of transaminase and lipid peroxidation, measured as thiobarbituric acid reactive substances [22]. In the previous research, authors also found that the liver induced-cirrhosis was associated with impairment in the pulmonary system, fluctuating gas exchange and pulmonary vessel size. That fact endorses the theory of endogenous and exogenous oxidative processes occurring in the pathology of cancer and also the affects to pulmonary capacities [23, 24].

The development of lung cancer has also been related with the formation of prooxidant species, especially those generated by phagocytes [13]. Some substances like proinflammatory cytokines (IL-1β, IL-6, and TNF-α) and the gene p53 have been related to the increase of ROS and other prooxidant molecules [25, 26].

The high levels of oxidative stress biomarkers found in breast cancer have been assigned to both disease progression and tissue damage, promoting the presence of ROS due to phagocyte activation [13]. Scientific literature has indicated that anticancer treatments can trigger ROS production, reinforcing the importance of checking oxidative stress biomarkers related to breast cancer [27]. The antioxidant capacity of serum seems to be dimished during breast cancer [28]. Similarly, the plasmatic values of uric acid, malonyl dialdehyde and nitric oxide have been shown to follow the same tendency in metastatic patients. All the three parameters decresed in patiens with metastasic cancer compared to healthy patients [28-30].

1.3.2. Oxidative Stress and Cardiovascular Diseases

The intake of diverse antioxidant vitamins can exert oxidative protection, leading to a reduction of the risk of atherosclerotic cardiovascular disease [31]. At functional levels, ROS is now able to exert its role as signaling molecules and therefore regulating an extensive variety of procedures in the cardiovascular system and thus contributing to the preservation of cardiovascular homeostasis [32]. In contrast, disproportionate increase of ROS generation exerts a crucial role in the initiation, progression and clinical consequences of cardiovascular diseases [33]. However, even though the great advancement in the comprehension of oxidative attacks in biology in particular and in cardiovascular medicine in general, the understanding of the essential mechanisms of cardiovascular diseases and consequences of unbalanced elevated ROS in cardiovascular tissue is still incomplete.

Oxidative stress is an important part in the development of atherosclerosis [34]. The oxidative attack to low-density lipoproteins leads to the uptaking of LDL by macrophages and the subsequent formation of foam cells. Moreover, the oxidative attack to lipids could develop in proinflammatory lipid peroxidation products [35]. Apolipoprotein E (apoE) is a class of apolipoprotein found in the chylomicron and intermediate-density lipoprotein (IDL) that is essential for the normal catabolism of triglyceride-rich lipoprotein constituents. Though expression of apoE is not able to reduce plasma cholesterol concentration, it can lead to a reversion of previous progressive atherosclerotic damages, leading to the reduction of foam cells.

Finally, the antiatherogenic capacity of apoE correlates with its antioxidant capacity *in vivo* [34].

The negative effects of ROS increment is comprised of not only the development of atherosclerosis, but experimental and clinical studies have showed the envelopment of ROS proliferation in the pathogenesis of atrial fibrillation [36], in addition to electrical heart remodelling [37]. The activity of the enzyme NADPH oxidase is boosted by activators of atrial fibrillation, for example angiotensin II and atrial stretch [37]. Finally, some compounds as N-acetylcysteine, polyunsaturated fatty acids or antioxidant vitamins have displayed the capability to reduce the oxidative attacks in the heart, while reducing cardiac arrhythmias [13].

1.3.3. Oxidative Stress and Neurodegenerative Diseases

The weakening of phospholipids from the cell membrane, as a main consequence of lipid peroxidation, has been characterized as the major origin of neurodegenerative diseases, such as Alzheimer disease or Parkinson disease [13, 38]. The accumulation of lipid peroxidation products and reduced enzymatic antioxidant protection are quite correlated with the formation of senile plaques and progression of Alzheimer disease [38]. Many established oxidative stress biomarkers such as acrolein, malondialdehyde or F_2-isoprostanes have been reported in Alzheimer brains and cerebrospinal fluid [13]. Hydroxynonenal have exerted high reactivity, hippocampal cytotoxicity, proving its accumulation in the brains of Alzheimer patients [39, 40]. The attack of ROS towards proteins leads to the formation of carbonyl groups. In patients with Alzheimer disease, high amounts of carvonyl groups were found in the frontal, parietal cortices and hippocampus of the brain [41, 42].

Mitochondrial damages triggered by the accumulation of β-amyloid in Alzheimer diasese cause the generation of ROS, resulting in cell damage and finally death [43, 44]. Due to the increase of oxidative stress, the concentration of polyunsaturated free fatty acids in the substantia nigra is depressed in Parkinson disease. Simultaneously, an increase can be detected in the lipid peroxidation biomarkers [45]. Moreover, elevated levels of 8-hydroxydeoxyguanosine were found, which is a product of an oxidative attack of guanosine that's associated with enhancement losses in the mitochondrial DNA [46].

1.3.4. Oxidative Stress and Diabetes

The oxidative injury produced by the increase of prooxidant species can disturb the normal function of the β-cell from pancreatic tissue [47, 48]. Ros is able to interact with the substrates involved in the insulin intracellular signaling, which leads to the cascade reactions that determine the type 2 diabetes mellitus complications concerning an augmented course in the polyols and hexosamines pathways [48, 49]. The high amounts of glucose boost the energy production in the cells, enhancing the reduction of NADH and $FADH_2$ in the electron transport chain occurring in the mitochondria. The continuous increase of the voltage gradient crossing the membrane of the mitochondria causes the onset of the blocking of complex III, leading to the reduction of coenzyme Q, molecular oxygen and superoxide anion ultimately [50].

Therefore, diabetic patients with deficient glycemic control, even treated with pharmaceutical drugs, are exposed to the increase of superoxide anion due to the enhancement in the oxidation of NADPH [13]. This fact has been demonstrated in the scientific literature, showing an increase of lipi peroxidation in diabetic patients [13, 48].

1.3.5. Oxidative Stress and Inflammatory Diseases

The increase of oxidative stress can lead to the development of diverse inflammatory responses, leading to diseases such as asthma and allergic rhinitis [47, 49-52]. Apart from lipid peroxidation, occurring in the nasal and airway mucosas, some other reactions have occurred due to the increase of oxidative stress. ROS production leads to a disturbance in cell signaling and arachidonic acid metabolism, enhancing airway and systemic inflammation [13]. Increased airway reactivity, chemoattack molecules, airway reactivity or increased vascular permeability can also be a consequence of ROS production [52].

The increase of allergic diseases can be promoted by the production of TH_1 and TH_2 cytokines, in fact that can be endorsed by the increase of oxidative stress, as well. Moreover, oxidative stress produces an overexpression of genes regulating inflammatory responses, such as alterations of TH_2 phenotype [53].

Diverse scientific studies have focused on the role of oxidative stress in the development and progression of asthma [54-56]. Asthma can be promoted due to oxidative stress in many ways as in the deletion of $GSTM_1$ and $GSTT_1$ [54] or methylation of the $ACSL_3$ gene in children [55]. Another focused disease is allergic rhinitis, which is supposed to be involved in the same way that asthma is [57]. The high levels of polyunsaturated fatty acids in the membranes of erythrocytes lead to increased susceptibility to oxidative injury. In fact, patients with allergic rhinitis present enhanced lipid peroxidation and increased superoxide dismutase enzyme activity at the level of erythrocytes [52, 58]. Increases in ceruloplasmin levels observed in patients suffering from allergic rhinitis are a consequence of the inflammation of airways. Moreover, ceruloplasmin is produced after an increased antioxidant response arising from cell injury [52].

REFERENCES

[1] Fang, Yun-Zhong, Sheng Yang, Guoyao Wu. 2002. "Free radicals, antioxidants, and nutrition." *Nutrition* no. 18 (10):872-879.

[2] Sun, Jie, Yi-Fang Chu, Xianzhong Wu, Rui Hai Liu. 2002. "Antioxidant and antiproliferative activities of common fruits." *Journal of Agricultural and Food Chemistry* no. 50 (25):7449-7454.

[3] Halliwell, Barry, John MC Gutteridge. 2015. *Free radicals in biology and medicine*: Oxford University Press, USA.

[4] Godic, Aleksandar, Borut Poljšak, Metka Adamic, Raja Dahmane. 2014. "The role of antioxidants in skin cancer prevention and treatment." *Oxidative medicine and cellular longevity* no. 2014.

[5] Inoue, Masayasu, Eisuke F Sato, Manabu Nishikawa, Ah-Mee Park, Yukimi Kira, Isuke Imada, Kozo Utsumi. 2003. "Mitochondrial generation of reactive oxygen species and its role in aerobic life." *Current Medicinal Chemistry* no. 10 (23):2495-2505.

[6] Valko, Marian, CJ Rhodes, Jan Moncol, MM Izakovic, Milan Mazur. 2006. "Free radicals, metals and antioxidants in oxidative stress-induced cancer." *Chemico-biological interactions* no. 160 (1):1-40.

[7] Guest, Jade, Ross Grant, Trevor A Mori, Kevin D Croft. 2014. "Changes in oxidative damage, inflammation and [NAD (H)] with age in cerebrospinal fluid." *PloS one* no. 9 (1):e85335.

[8] Hernandez, Angel Gil DRT. 2010. *Tratado de nutrición: Nutrición clínica*. Vol. 4: Ed. Médica Panamericana.

[9] Murphy, Robert C, Simona Zarini. 2002. "Glutathione adducts of oxyeicosanoids." *Prostaglandins & other lipid mediators* no. 68:471-482.

[10] Liu, Jiankang, Helen C Yeo, Eva Övervik-Douki, Tory Hagen, Stephanie J Doniger, Daniel W Chu, George A Brooks, Bruce N Ames. 2000. "Chronically and acutely exercised rats: biomarkers of oxidative stress and endogenous antioxidants." *Journal of Applied Physiology* no. 89 (1):21-28.

[11] Halliwell, B, JMC Gutteridge. 2007. Free radicals in biology and medicine, 4th edn. Clarendon. Oxford.

[12] Halliwell B, Gutteridge JMC. 2007. *Free radicals in biology and medicine*. 4th ed: ed. Oxford: Oxford university press.

[13] Pisoschi, Aurelia Magdalena, Aneta Pop. 2015. "The role of antioxidants in the chemistry of oxidative stress: a review." *European journal of medicinal chemistry* no. 97:55-74.

[14] Yin, Huiyong, Libin Xu, Ned A Porter. 2011. "Free radical lipid peroxidation: mechanisms and analysis." *Chemical Reviews* no. 111 (10):5944-5972.

[15] Martysiak-Żurowska, Dorota, Weronika Wenta. 2012. "A comparison of ABTS and DPPH methods for assessing the total antioxidant capacity of human milk." *Acta Sci Pol Technol Aliment* no. 1 (11):83-89.

[16] Tandon, Vishal R, Sudhaa Sharma, Annil Mahajan, Ghulam Hussain Bardi. 2005. "Oxidative stress: a novel strategy in cancer treatment." *JK Science* no. 7 (1):1-3.

[17] Raafat, Yousri, Noaman Eman, Shawi Omama El, Fahmy Nadia, Ghazy Maha. 2011. "Evaluation of Anti-Oxidant Status and Radioprotective Activity of a Novel Anti-Cancer Drug in Mice." *Journal of Cancer Therapy* no. 2011.

[18] Szabo, Eva, Maura E Riffe, Seth M Steinberg, Michael J Birrer, R Ilona Linnoila. 1996. "Altered cJUN expression: an early event in human lung carcinogenesis." *Cancer Research* no. 56 (2):305-315.

[19] Stehbens, William E. 2004. "Oxidative stress in viral hepatitis and AIDS." *Experimental and Molecular Pathology* no. 77 (2):121-132. doi: http://dx.doi.org/10.1016/j.yexmp.2004.04.007.

[20] Roberts Ii, L. J., K. P. Moore, W. E. Zackert, J. A. Oates, J. D. Morrow. 1996. "Identification of the major urinary metabolite of the F2-isoprostane 8- iso-prostaglandin F(2α) in humans." *Journal of Biological Chemistry* no. 271 (34):20617-20620. doi: 10.1074/jbc.271.34.20617.

[21] Boveris, A, OR Koch, AOM Stoppani. 1978. "Decreased rate of hydrogen-peroxide production by liver-mitochondria in chronic experimental alcoholism." *Medicina-Buenos Aires* no. 38 (6):647-651.

[22] Salatti Ferrari, Renata, Darlan Pase da Rosa, Luiz Felipe Forgiarini, Silvia Bona, Alexandre Simões Dias, Norma Possa Marroni. 2012. "Oxidative stress and pulmonary changes in experimental liver cirrhosis." *Oxidative medicine and cellular longevity* no. 2012.

[23] Park, Tae Jun, Ji Yeon Kim, S Paul Oh, So Young Kang, Bong Wan Kim, Hee Jung Wang, Kye Yong Song, Hyoung Chin Kim, In Kyoung Lim. 2008. "TIS21 negatively regulates hepatocarcinogenesis by disruption of cyclin B1–Forkhead box M1 regulation loop." *Hepatology* no. 47 (5):1533-1543.

[24] Rajendrasozhan, Saravanan, Se-Ran Yang, Indika Edirisinghe, Hongwei Yao, David Adenuga, Irfan Rahman. 2008. "Deacetylases and NF-κ B in Redox Regulation of Cigarette Smoke-Induced Lung Inflammation: Epigenetics in Pathogenesis of COPD." *Antioxidants & redox signaling* no. 10 (4):799-812.

[25] Stewart, Zoe A, Jennifer A Pietenpol. 2001. "p53 Signaling and cell cycle checkpoints." *Chemical Research in Toxicology* no. 14 (3):243-263.

[26] Srinivas, A, P Jaganmohan Rao, G Selvam, A Goparaju, Balakrishna P Murthy, Neelakanta P Reddy. 2012. "Oxidative stress and inflammatory responses of rat following acute inhalation exposure to iron oxide nanoparticles." *Human & experimental toxicology* no. 31 (11):1113-1131.

[27] Conklin, Kenneth A. 2004. "Chemotherapy-associated oxidative stress: impact on chemotherapeutic effectiveness." *Integrative cancer therapies* no. 3 (4):294-300.

[28] Omar ME, Abdel-Salam, Youness Eman R, Hafez Hafez F. 2011. "The antioxidant status of the plasma in patients with breast cancer undergoing chemotherapy." *Open Journal of Molecular and Integrative Physiology* no. 2011.

[29] Ruiz-Larrea, Ma Begoña, Ana Ma Leal, Mariana Liza, Mercedes Lacort, Herbert de Groot. 1994. "Antioxidant effects of estradiol and 2-hydroxyestradiol on iron-induced lipid peroxidation of rat liver microsomes." *Steroids* no. 59 (6):383-388.

[30] Koracevic, D, G Koracevic, V Djordjevic, S Andrejevic, V Cosic. 2001. "Method for the measurement of antioxidant activity in human fluids." *Journal of clinical pathology* no. 54 (5):356-361.

[31] Rimm, Eric B., Meir J. Stampfer. 2000. "Antioxidants for vascular disease." *Medical Clinics of North America* no. 84 (1):239-249. doi: http://dx.doi.org/10.1016/S0025-7125(05)70216-9.

[32] Dröge, Wulf. 2002. "Free radicals in the physiological control of cell function." *Physiological reviews* no. 82 (1):47-95.

[33] Lakshmi, S Vijaya, G Padmaja, Periannan Kuppusamy, Vijay Kumar Kutala. 2009. "Oxidative stress in cardiovascular disease." *Indian J Biochem Biophys* no. 46 (6):421-440.

[34] Meagher, Emma, Daniel J. Rader. 2001. "Antioxidant Therapy and Atherosclerosis: Animal and Human Studies." *Trends in Cardiovascular Medicine* no. 11 (3–4):162-165. doi: http://dx.doi.org/10.1016/S1050-1738(01)00105-0.

[35] Witztum, Joseph L, Daniel Steinberg. 1991. "Role of oxidized low density lipoprotein in atherogenesis." *Journal of Clinical Investigation* no. 88 (6):1785.

[36] Ali-Hassan-Sayegh, Sadegh, Seyed Jalil Mirhosseini, Mohammad Rezaeisadrabadi, Hamid Reza Dehghan, Farbod Sedaghat-Hamedani, Elham Kayvanpour, Aron-Frederik Popov, Oliver J Liakopoulos. 2014. "Antioxidant supplementations for prevention of atrial fibrillation after cardiac surgery: an updated comprehensive systematic review and meta-analysis of 23 randomized controlled trials." *Interactive cardiovascular and thoracic surgery*:ivu020.

[37] Berliner, Judith A, Mohamad Navab, Alan M Fogelman, Joy S Frank, Linda L Demer, Peter A Edwards, Andrew D Watson, Aldons J Lusis. 1995. "Atherosclerosis: basic mechanisms oxidation, inflammation, and genetics." *Circulation* no. 91 (9):2488-2496.

[38] Markesbery, William R. 1997. "Oxidative Stress Hypothesis in Alzheimer's Disease." *Free Radical Biology and Medicine* no. 23 (1):134-147. doi: http://dx.doi.org/10.1016/S0891-5849(96)00629-6.

[39] Lovell, Mark A, William D Ehmann, Steven M Butler, William R Markesbery. 1995. "Elevated thiobarbituric acid-reactive substances and antioxidant enzyme activity in the brain in Alzheimer's disease." *Neurology* no. 45 (8):1594-1601.

[40] Selley, M. L., D. R. Close, S. E. Stern. 2002. "The effect of increased concentrations of homocysteine on the concentration of (E)-4-hydroxy-2-nonenal in the plasma and cerebrospinal fluid of patients with Alzheimer's disease." *Neurobiology of Aging* no. 23 (3):383-388. doi: http://dx.doi.org/10.1016/S0197-4580(01)00327-X.

[41] Hensley, Kenneth, Nathan Hall, Ramachandran Subramaniam, Pamela Cole, Marni Harris, Michael Aksenov, Marina Aksenova, S Prasad Gabbita, Jun F Wu, John M Carney. 1995. "Brain regional correspondence between Alzheimer's disease histopathology and biomarkers of protein oxidation." *Journal of neurochemistry* no. 65 (5):2146-2156.

[42] Hensley, K., D. A. Butterfieldld, N. Hall, P. Cole, R. Subramaniam, R. Mark, M. P. Mattson, W. R. Markesbery, M. E. Harris, M. Aksenov, M. Aksenova, J. F. Wu, J. M. Carney. 1996. "Reactive Oxygen Species as Causal Agents in the Neurotoxicity of the Alzheimer's Disease-Associated Amyloid Beta Peptidea." *Annals of the New York Academy of Sciences* no. 786 (1):120-134. doi: 10.1111/j.1749-6632.1996.tb39057.x.

[43] Bobba, A., G. Amadoro, D. Valenti, V. Corsetti, R. Lassandro, A. Atlante. 2013. "Mitochondrial respiratory chain Complexes I and IV are impaired by β-amyloid via direct interaction and through Complex I-dependent ROS production, respectively." *Mitochondrion* no. 13 (4):298-311. doi: http://dx.doi.org/10.1016/j.mito.2013.03.008.

[44] Luque-Contreras, Diana, Karla Carvajal, Danira Toral-Rios, Diana Franco-Bocanegra, Victoria Campos-Peña. 2014. "Oxidative stress and metabolic syndrome: cause or consequence of Alzheimer's disease?" *Oxidative medicine and cellular longevity* no. 2014.

[45] Dalfó, Esther, Manuel Portero-Otín, Victoria Ayala, Anna Martínez, Reinald Pamplona, Isidre Ferrer. 2005. "Evidence of oxidative stress in the neocortex in incidental Lewy body disease." *Journal of Neuropathology & Experimental Neurology* no. 64 (9):816-830.

[46] Bender, Andreas, Kim J Krishnan, Christopher M Morris, Geoffrey A Taylor, Amy K Reeve, Robert H Perry, Evelyn Jaros, Joshua S Hersheson, Joanne Betts, Thomas Klopstock. 2006. "High levels of mitochondrial DNA deletions in substantia nigra neurons in aging and Parkinson disease." *Nature Genetics* no. 38 (5):515-517.

[47] Valko, Marian, Dieter Leibfritz, Jan Moncol, Mark T. D. Cronin, Milan Mazur, Joshua Telser. 2007. "Free radicals and antioxidants in normal physiological functions and human disease." *The International Journal of Biochemistry & Cell Biology* no. 39 (1):44-84. doi: http://dx.doi.org/10.1016/j.biocel.2006.07.001.

[48] Bandeira, Suziy de M, Glaucevane da S Guedes, Lucas José S da Fonseca, André S Pires, Daniel P Gelain, José Claudio F Moreira, Luíza A Rabelo, Sandra Mary L Vasconcelos, Marília Oliveira F Goulart. 2012. "Characterization of blood oxidative stress in type 2 diabetes mellitus patients: increase in lipid peroxidation and SOD activity." *Oxidative medicine and cellular longevity* no. 2012.

[49] Evans, Jonathan St BT, David E Over, Simon J Handley. 2005. "Suppositions, extensionality, and conditionals: A critique of the mental model theory of Johnson-Laird and Byrne (2002)."

[50] Brownlee, Michael. 2005. "The pathobiology of diabetic complications a unifying mechanism." *Diabetes* no. 54 (6):1615-1625.

[51] Marple, Bradley F. 2010. "Allergic rhinitis and inflammatory airway disease: interactions within the unified airspace." *American journal of rhinology & allergy* no. 24 (4):249-254.

[52] Sequeira, Shiefa, Ashalatha V Rao, Anjali Rao. 2012. "Increased oxidative stress and altered antioxidants status in patients with chronic allergic rhinitis." *Advances in Bioscience and Biotechnology* no. 3 (7A):951.

[53] King, Miranda R, Anisa S Ismail, Laurie S Davis, David R Karp. 2006. "Oxidative stress promotes polarization of human T cell differentiation toward a T helper 2 phenotype." *The Journal of Immunology* no. 176 (5):2765-2772.

[54] Liang, Siqiao, Xuan Wei, Chen Gong, Jinmei Wei, Zhangrong Chen, Xiaoli Chen, Zhibo Wang, Jingmin Deng. 2013. "Significant association between asthma risk and the GSTM1 and GSTT1 deletion polymorphisms: An updated meta-analysis of case–control studies." *Respirology* no. 18 (5):774-783. doi: 10.1111/resp.12097.

[55] Perera, Frederica, Wan-yee Tang, Julie Herbstman, Deliang Tang, Linda Levin, Rachel Miller, Shuk-mei Ho. 2009. "Relation of DNA methylation of 5′-CpG island of ACSL3 to transplacental exposure to airborne polycyclic aromatic hydrocarbons and childhood asthma." *PloS one* no. 4 (2):e4488.

[56] Yamamoto, Masatsugu, Amrit Singh, Francesco Sava, Mandy Pui, Scott J Tebbutt, Christopher Carlsten. 2013. "MicroRNA expression in response to controlled exposure to diesel exhaust: attenuation by the antioxidant N-acetylcysteine in a randomized crossover study." *Environmental Health Perspectives (Online)* no. 121 (6):670.

[57] Bowler, Russell P, James D Crapo. 2002. "Oxidative stress in allergic respiratory diseases." *Journal of Allergy and Clinical Immunology* no. 110 (3):349-356.

[58] Aguirre, F., I. Martin, D. Grinspon, M. Ruiz, A. Hager, T. De Paoli, J. Ihlo, H. A. Farach, C. P. Poole Jr. 1998. "Oxidative Damage, Plasma Antioxidant Capacity, and Glucemic Control in Elderly NIDDM Patients." *Free Radical Biology and Medicine* no. 24 (4):580-585. doi: http://dx.doi.org/10.1016/S0891-5849(97)00293-1.

Chapter 2

FORMATION OF PROSTANOIDS

Débora Villaño Valencia, PhD
and Begoña Cerdá, PhD

ABSTRACT

Eicosanoids are a group of molecules originating from lipid character oxidation of essential fatty acids of 20 carbon atoms. All eicosanoids are molecules of 20 carbon atoms and are grouped into prostaglandins, thromboxanes, leukotrienes, hydroxy acids and certain leukotriene precursors. They're molecules involved in complex networks of cellular communication in animal organisms, including humans.

Prostanoids belong to the eicosanoids group, and are hydroxylated derivatives of polyunsaturated fatty acids (PUFA); PUFA precursors are restricted to the family of essential fatty acids (EFAs) w6 and w3. The eicosanoids are synthesized *in vivo* by different metabolic pathways, and some compounds are formed by more than one enzymatic mechanism. The main pathways mediated include:

- PGHS or cyclooxygenase (COX): which generates Prostaglandin H (PGH), cyclic product, a precursor of prostanoids formed by the cicloxigenación and sequential hydroperoxidation of PUFA.
- Lipoxygenase (LOX): this pathway generates linear products, initiated by the action of the enzyme 5-lipoxygenase (5-LOX), which are collectively referred to as leukotrienes (LT) and include substances of delayed hypersensitivity, cysteinyl leukotrienes LTC4 , LTD4, and LTE4, actively synthesized

during anaphylactic shock reaction. Other products, such as hydroperoxy and hydroxyeicosatetraenoic acids are generated by the 12- or 15-LOX while lipoxins, negative modulatory molecules of the inflammatory process, appear in the sequential action of more than one LOX.

- Epoxygenase: this pathway generates mono-oxygenated products formed by hydroxylation of PUFA on CH3-terminal, C-adjacent on or at the expense of a double bond in the form of epoxides catalyzed by cytochrome P450. The endothelium-derived hyperpolarizing factor (EDHF), is one of the products generated in this way.

Eicosanoids produce a wide range of biological effects in inflammatory responses (predominantly in the joints, skin, and eyes), duration and intensity of pain and fever, and reproductive function (including induction of labor). Also they fulfill important functions inhibiting gastric acid secretion, regulating blood pressure through vasoconstriction and vasodilatation, and inhibiting or activating platelet aggregation and thrombosis.

Keywords: cyclooxygenase, lipoxygenase, arachidonic acid, prostanoids, synthesis

INTRODUCTION

Lipids are a heterogeneous group of natural substances that include fats, monoglycerides, diglycerides and triglycerides, certain vitamins and hormones, waxes and sterols, among others. Their function in the body is related to energy store and they are the main components of cell membrane. Some compounds such as sphingolipids or steroid hormones are of lipid nature and act as signaling molecules. Others like prostaglandins play a key role in the progression of inflammatory conditions [1].

Prostaglandins (PGs) were first identified in human seminal fluid by Ulf Von Euler in the early 1930s [2], who thought they were originated by prostate gland (hence their name). In fact they are produced in all mammalian cells (except blood cells). They belong to the family of eicosanoids as they possess twenty carbon atoms (in Greek eikosi = twenty) as well as the related compounds prostacyclins, thromboxanes and leukotrienes.

These eicosanoids mediate in important physiological functions, as the inflammatory response, regulation of blood pressure or blood clotting, by means of binding to G-protein-coupled receptors and activating intracellular signal transduction of proteins involved in these processes [3].

The main precursor of prostaglandins in humans is arachidonic acid (5,8,11,14-eicosatetraenoic acid), a C20 fatty acid with four non-conjugated double bonds. It is synthesized from the essential fatty acid, the linoleic acid. Arachidonic acid also leads to the formation of leukotrienes. The metabolism of eicosanoids also renders prostacyclins and thromboxanes. The specific metabolic pathway depends on the tissue involved. Blood platelets produce thromboxanes almost exclusively while vascular endothelial cells mainly produce prostacyclins.

Prostaglandins contain a cyclopentane ring with two side chains oriented in trans relative to the ring. The allocated letter (A, B, C, ecc) refers to different functional groups allocated in carbons 9 and 11 of the cyclopentane ring. The subscript number indicates the number of double bonds contained on the side chain, i.e., PGA1 is an α,β-unsaturated ketone with one double bond [4].

Thromboxanes have an oxygenated six-member ring as well as two side chains and a similar nomenclature which has been created for the location of the double bonds. Leukotrienes have three conjugated double bonds and the subscript indicates the total number of double bonds in the molecule [4].

Both prostaglandins (PG) and thromboxanes (TX) are products of the cyclooxygenase pathway and leukotrienes (LT) are products of the 5-lipooxygenase pathway.

Eicosanoids are usually associated with inflammatory response, however there are some products with anti-inflammatory properties. An example is lipoxins (LXs), products of 12- and 15 lipooxygenase pathways, which seem to inhibit leukotriene actions. At present two lipoxins have been identified; lipoxin A4 (LXA4) and lipoxin B4 (LXB4).

In this chapter we will focus on the biosynthetic pathways of these compounds.

2.1. Enzymatic Oxidation of Fatty Acids

The majority of the fatty acids on the organism are bound to glycerol, leading to triglycerides which are the main reserves of energy in the human organism. Triglycerides are formed by one molecule of glycerol esterified with three molecules of fatty acid [5].

They are released by lipolysis from the adipose tissue in fatty acids and may be used by the tissue as an energy source. In addition, they can also be esterified obtaining new triglycerides, or they can be accumulated and spread through the plasma bound to albumin, acting as an energy source in peripheral tissues. Fatty acids are highly energetic metabolic molecules. But their importance lies not only in their energy capacity, but can also in that they serve as essential constituents of natural lipids such as esterified cholesterol, phospholipids or glycolipids being major components of biomembranes[5].

Several pathways for the oxidation of fatty acid in living organisms have been observed. The most important and well-known pathway for the oxidation of fatty acids is β-oxidation, which leads to the conversion of fatty acids into acetyl-CoA or propionyl-CoA (derived from odd-numbered fatty acid chains). Finally, those CoAs are oxidized into carbon dioxide and H_2O resulting in energy production [5, 6].

There are two other types of oxidations:

- α-oxidation: involves hydroxylation of α-carbon, and the removal of the fatty acid carboxyl terminus and conversion of the α-hydroxyl group in carboxyl group. Once the compound generated binds coenzyme A, the ß-oxidation can then be entered.
- ω-oxidation: The enzymes carrying out this process are localized in the endoplasmic reticulum liver and kidney and act on fatty acids (preferably with 10-12 carbons). Failures affecting the ß-oxidation may promote a higher rate of ω-oxidation.

The β-oxidation takes place in the second carbon counting from the (β carbon) carbonyl group, occuring in both mitochondria and peroxisomes. Generally, both models differ in metabolic fluxes. Mitochondrial β-oxidation is quite efficient, usually converting R-CoA to the final product acetyl-CoA. This pathway constitutes the major process by which fatty acids are oxidized to produce energy. Peroxisomal β-oxidation does not proceed via channelization, and its intermediates may be accumulated in cells [7].

For the correct development of β-oxidation, the fatty acids have to be released from the stored triglycerides, being regulated by hormonal mechanisms. Adrenaline and glucagon are secreted due to a descent of the plasmatic glucose levels, activating the mobilization of the stored triglycerides in the adipose tissue. Briefly, triglycerides are transported to the tissues where they are oxidized producing energy. This hormone secretion causes the activation of adenylate cyclase located in the membrane of adipocytes. That activation leads to an increase of ciclyc adenosin monophosphate (cAMP), which binds to protein kinase catalyzing the hydrolysis of the esterified triglycerides leading to fatty acids. Finally, they are diffused into the blood, joining to albumin. The β-Oxidation has no place neither in brain cells nor red cells; therefore, the glucolysis is the only way to obatin energy in such cells [8].

Once the fatty acids have been released they have to be activated in order to to enter the mitochondria, undergoing a series of enzymatic reactions. The first reaction is catalyzed by acyl-CoA synthetase (along the length of the fatty acids involved in different isoenzymes), resulting in the generation of acyl-CoA (activated fatty acid). The acyl-CoA binds to carnitine and it is then transported into the mitochondria. Short and medium fatty acids (less than 12 carbon atoms) do not need the transport chain of carnitine, crossing the mitochondrial membrane directly. In the layer of the inner mitochondrial membrane, there is acyltransferase carnitine 1 (CAT-1). CAT-1 from the outer mitochondrial membrane that eliminates the coenzyme A molecule acyl-CoA. Simultaneously CAT-1 joins the carnitine located in the intermembrane space originating acylcarnitine. The CoA is released into the cytosol to activate another fatty acid. Briefly, a carrier protein called translocase transfers acylcarnitine to the mitoncondrial matrix. Meanwhile, the acyltransferase carnitine 2 (CAT-2) located on the inner side of the inner mitochondrial membrane, binds a molecule of CoA to the fatty acid regenerating the acyl-CoA. Carnitine is returned to the intermembrane space by the carrier protein and reacts with other acyl-CoA, repeating the cycle. The acyl- CoA is then prepared to suffer the β-oxidation [9, 10]. The β-oxidation is regulated by the availability of fatty acids and malonyl-CoA (inhibiting CAT-1). This process of β-oxidation occurs in different steps [5]:

a) Formation of Coenzyme A thioester.
b) α,β-dehydrogenation of the acyl-CoA.
c) Hydration of α,β-unsaturated fatty acyl-CoA.

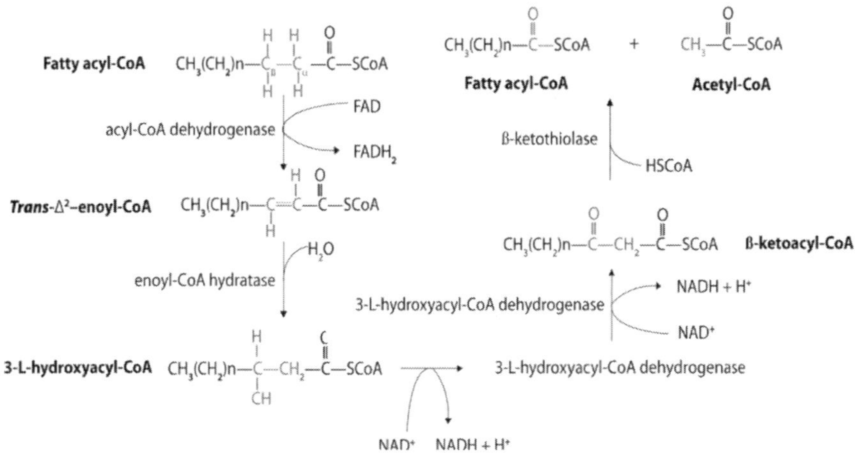

Figure 2.1. β-oxidation of fatty acid in mitochondria.

d) Oxidation of the β-hydroxyacyl CoA.
e) Thiolytic cleavage of the β-ketoacyl-CoA.

Fatty acids penetrate the mitochondria, being linked to coenzyme A in the outer mitochondrial membrane. Oxidative reactions of the β-oxidation cycle are preceded by the activation of the fatty acid to its thioester with coenzyme A (CoA) catalyzed by ATP-dependent ligase. As a result of dehydrogenation of the thioester the enoyl-CoA is formed. After its hydration, owing to enoyl-CoA hydratase, the hydroxyacyl-CoA is then formed. The dehydrogenation reactions and the addition of a water molecule to the trans-Δ^2-enoyl-CoA are stereoselective. The third reaction of this pathway is the oxidation of the hydroxyl group, catalyzed by 3-hydroxyacyl-CoA dehydrogenase. The thiolase catalyzes the thiolytic cleavage of β-ketoacyl-CoA into two molecules of acyl-CoA as end products [9, 10].

The final result of the β-oxidation is a fatty acid with two carbon atoms less and a molecule of acetyl-CoA (Figure 2.1). The process is repeated until finally all the fatty acid is broken into units of acetyl CoA. However, in the case of fatty acids of even numbers of carbon atoms the final product of β-oxidation is malonyl-CoA [5].

This reaction sequence takes place for saturated fatty acids. Unsaturated fatty acids undergo a series of further reactions allowing movement of the double bonds, so that they are liable to attack by the enzymes of β-oxidation [5, 8]. In the case of unsaturated fatty acids the intervention of three new enzymes is also required. The first enzyme is enoyl CoA isomerase, catalyzing the conversion of the double bond to trans-Δ^2 cis-Δ^3 form, which is the natural substrate CoA-hydratase-enoyl. Secondly, The enzyme 2,4-dienoyl-CoA reductase reduces the double bond giving trans-3-enoyl CoA which is isomerized to trans-2-enoyl CoA by 3,2-enoyl CoA-isomerase [8].

2.2. SYNTHESIS OF PROSTAGLANDINS

Prostaglandins (PGs) are bioactive lipids derived from plasma and nuclear membranes that are produced by the action of cyclooxygenase (COX) and further specific enzymes in the arachidonic acid metabolic pathway. PGs are metabolites of arachidonic acid, a 20-carbon tetraenoic fatty acid (C20:4ω6). Arachidonic acid is a fatty acid that enters the body from meat or its precursor, linoleic acid (C18:2) found in vegetables. Arachidonic acid is absorbed in the intestine and circulates bound to albumin throght the organism. Moreover, it could be synthesized from linolenic acid (Figure 2.2) due to enzymatic patway [11].

PGs are found in most tissues and organs and are produced in the majority of the nucleated cells. They are synthesized *de novo* when their action is needed and they are rapidly degraded after exerting their effect. PGs are autocrine and paracrine lipid mediators that act on platelets, endothelium, uterine cells and mast cells, among others [11]. The name prostaglandin comes from the prostate gland. They was firstly isolated in 1930s in the seminal fluid, prostate, and seminal vesicles by Von Euler (1936), showing decreased blood pressure and increased smooth contraction. PGs exert variated and important actions in the organism, including [12-22]:

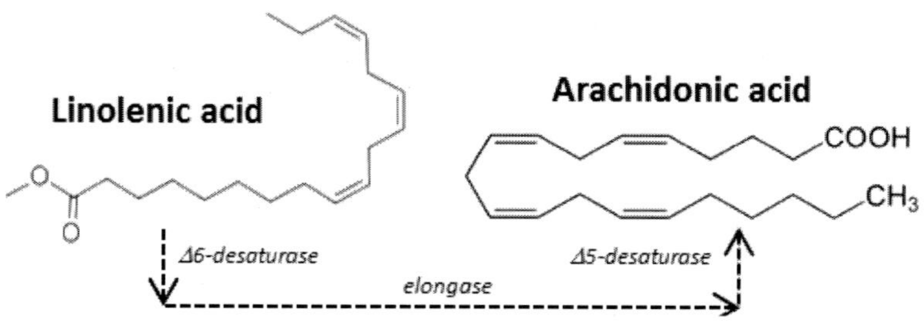

Figure 2.2. Formation of arachidonic acid from linolenic acid.

- PGs are involved in the inflammatory response of tissues: vasodilation, increased tissue permeability allowing passage of leukocytes, antiplatelet, or stimulation of the nerve endings of pain.
- Increased secretion of gastric mucus, and decreased gastric acid secretion.
- They cause a contraction of smooth muscles. This is especially important in the woman's uterus. Similarly, they are released during menstruation in favor the detachment of endometrium. In human semen, small amounts of prostaglandins are released to promote the contraction of the uterus and consequently the ascension of sperm to the fallopian tubes.
- They're involved in the regulation of body temperature.
- They control the decrease of blood pressure in favor the elimination of substances in the kidney.

Prostaglandins bind to a subfamily of transmembrane cell surface receptors, these coupled receptors are called G-protein DP_{1-2}, EP_{1-4}, FP, IP, and TP, referring to the corresponding receptor links prostaglandin (eg DP_{1-2} PGD_2 receptors bind) [23].

As previously mentioned, arachidonic acid is the precursor in the synthesis of prostanoids. These products of arachidonic acid metabolism are together known as eicosanoids, due to "eikosi" (the Greek word for twenty) which is the number of carbon atoms in arachidonic acid. Arachidonic acid is widely distributed throughout the organism, being a constituent of cell membranes. Once released from the membrane, the AA can follow three different routes:

1. Cyclooxygenase pathway, which will leads to prostanoids, including prostaglandins, prostacyclins and thromboxanes.
2. Lipooxigenase pathway, leading to leukotrienes.
3. Cytochrome P450 pathway, leading to epoxyeicosatrienoic acids and their metabolites.

2.2.1. Cyclooxygenase Pathway

PGs and thromboxane A_2 (TXA_2) are formed when arachidonic acid is released from the plasma membrane by phospholipases (PLAs) and metabolized by the sequential action of prostaglandin G/H synthase, cyclooxygenase (COX), and respective synthases [24]. Phospholipase A_2 (PLA_2), in particular cytoplasmic phospholipase A_2 ($cPLA_2$), is the first enzyme engaged in synthesizing these compounds, as it specifically releases arachidonic acid from lipids in the cell membrane, which then enters cyclooxygenase pathways [25].

PGs production depends on the activity of cyclooxygenases (COX), bifunctional enzymes that exert both cyclooxygenase and peroxidase activity. Three isoforms of COX have been identified [26-28]. However, the two main COX isoforms are cyclooxygenase-1 (COX-1) and cyclooxygenase-2 (COX-2) (Figure 2.3).

COX-1 is constitutively expressed in most tissues and is responsible for synthesising PGs with physiological function: vascular homeostasis, protecting gastric mucosa and maintaining renal integrity. COX-2, in contrast to COX-1, is barely detectable under physiological conditions. The expression of COX-2 is regulated by pro-inflammatory stimuli: growth factors and inflammatory responses. Isoenzyme COX-3 is derived from the gene encoding the COX-1, but is structurally and functionally different. The main action of COX-3 is the production of a fever and pain at the central level and not at the level of inflammation [29, 30].

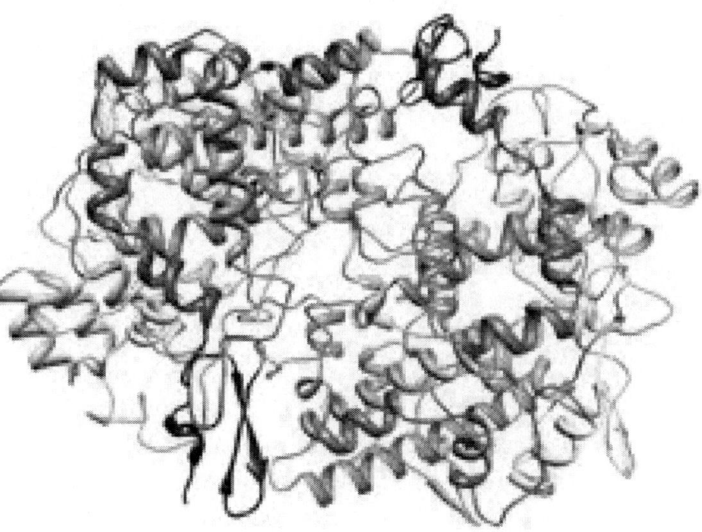

Figure 2.3. Cyclooxygenase-2.

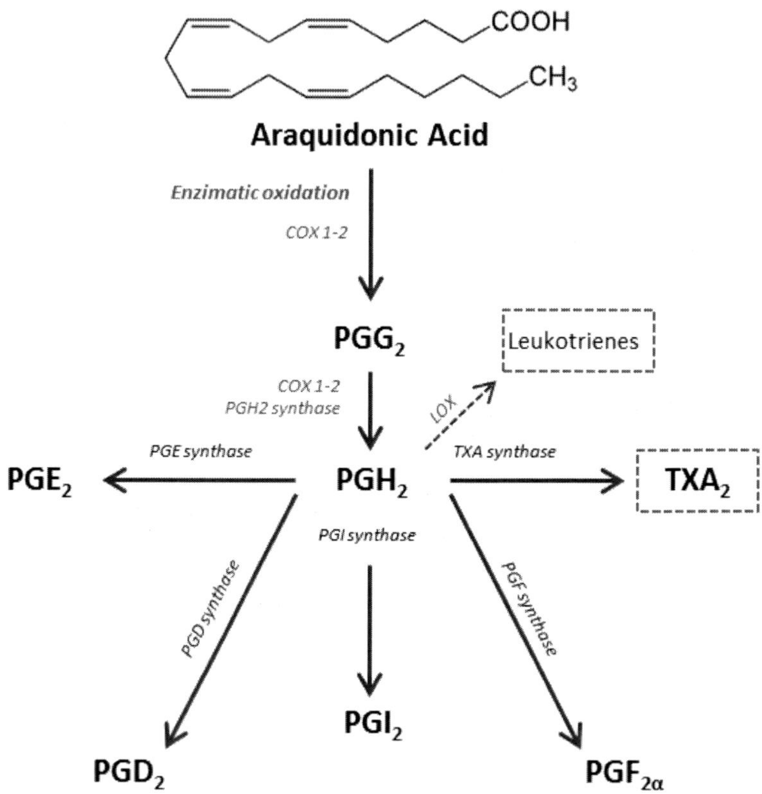

Figure 2.4. Formation of prostanoids and leukotrienes.

The eicosanoid cascade starts with the activation of PLA_2 and PLAC that release arachidonic acid from the cellular membrane. Arachidonic acid is oxidized and then reduced by the enzymes COX-1 and COX-2. The resulting metabolite is PGG_2, a cyclopentane hydroperoxy endoperoxide; the peroxidase activity of COXs then reduces PGG_2 to PGH_2. Both reactions are functionally interconnected. COXs are key enzymes in PG biosynthesis and differ in their expression levels and tissue distribution. The COX-1 gene is principally constitutive in function and possesses a typical GC-rich housekeeping promoter. In contrast, the COX-2 gene resembles an early response gene. It is strongly induced by mitogenic and proinflammatory stimuli, promoted by inhibitors of protein synthesis, and acutely regulated at both transcriptional and posttranscriptional levels [29, 30].

The mechanism by which arachidonic acid is converted to PGH_2 has been proposed by Van der donk in 2002 and Rouzer in 2003 [31, 32]. The newly synthesized COX enzyme needs to be activated to produce a tyrosyl radical. This tyrosyl radical converts arachidonic acid to an arachidonyl radical, which reacts with two molecules of oxygen to yield PGG_2. PGG_2 diffuses into the peroxidase site and is reduced to PGH_2 by the peroxidase activity (Figure 2.4). The cyclooxygenase activity is dependent on heme oxidation - that is, on the peroxidase activity - but continuous peroxidase activity is not necessary for cyclooxygenase activity, as the tyrosyl radical is regenerated in each catalytic cycle. PGH_2 is the root prostaglandin from which prostaglandin isomers such as thromboxane and prostacyclin are made by downstream synthases, via isomerization and oxidation or reduction reactions. Cyclooxygenases have short catalytic lifespans because the enzyme is autoinactivated. The mechanism of autoinactivation is unknown, but reactive tyrosyl radicals may cause internal protein modification [30].

When COXs enzymes transform arachidonic acid into PGH_2, this is then again transformed by cell-specific synthases, mainly cytosolic prostaglandin E synthase (cPGES), microsomal prostaglandin E synthase-1 (mPGES-1), microsomal prostaglandin E synthase-2 (mPGES-2), prostaglandin I synthase (PGIS), and thromboxane synthase (TxS) into the five principal prostaglandins (PGs) and thromboxanes [30, 32].

2.2.2. Synthesis of Thromboxanes

Polyunsaturated fatty acids (PUFA) serve as substrates for cyclooxygenase enzymes (COX). Arachidonic acid is the major polyunsaturated fatty acid in membrane cells and is a substrate for thromboxane synthesis pathway. Arachidonic acid is esterified in membranes until it's released by phospholipase A_2, when cells are activated by mechanical trauma in membranes or by chemical stimuli such as cytokines or growth factors, among others [33].

As mentioned earlier, COX-I catalyses the conversion of arachidonic acid into PGG_2, which is the substrate of a hydroxyperoxidase. The product, which is the stable hydroperoxide PGH_2. PGH_2, is then converted into thromboxane A_2 by a specific thromboxane synthase (TXAS).

Thromboxane A_2 is a 2-series prostanoid, containing 2 double bonds making it a potent mediator of platelet aggregation. It is released during vascular inflammation, for example in atherosclerosis or vascular injury. TXA_2 interacts with its thromboxane prostanoid (TP) cell membrane receptor coupled to G-protein. The interaction is followed by a signal transduction cascade in the target cells, like platelets, endothelial cells and vascular smooth muscle cells (Gabrielsen et al., 2010).

2.2.3. Synthesis of Leukotrienes

Leukotrienes belong to the eicosanoid family as they are twenty carbon fatty acid derivatives, generated from the essential arachidonic acid (20:4ω6) by way of the 5-lypooxygenase enzyme (5-LOX). They are 4-series prostanoids, as leukotriene B_4, containing 4 double bonds.

The 5-lipooxygenase belongs to the family of lipooxygenases (LOXs) that catalyse the oxygenation of polyunsaturated fatty acids, such as arachidonic acid or linoleic acid. The addition of oxygen to AA produces hydroxyperoxyeicosatetraenoic acid (HPETE), which is further transformed to give leukotrienes (LTs) (Schweiger et al., 2007). Each isoform of lipooxygenase has unique properties like substrate specificity (Mashima & Okuyama, 2015).

Leukotrienes are mainly produced in the cells of the immune system, such as polymorphonuclear leukocytes, macrophages or mast cells. Different stimuli like immune complexes of immunoglobulin E (IgE)-antigen or bacterial components activate the translocation of the enzyme 5-lipooxygenase to the nuclear membrane. The translocation seems to be favoured by specific binding regions in the enzyme, with SH-groups that interact with proteins, as seen in the 5-lipooxygenase-activating protein (FLAP). FLAP is a protein in the nuclear membrane that facilitates the interaction of 5-LOX with its substrate, the arachidonic acid. Currently, the current developments of pharma companies for asthma control are inhibitors for 5-LOX activating protein FLAP, leading to suppression of LT production (Petersen et al., 2015).

It has been proposed that the 5-LOX possesses a NH_2-terminal domain that binds two calcium ions in a dative or coordinated link, as well as a large catalytic domain that binds iron (Gilbert et al., 2011). Metal binding allows the enzyme to catalyse the conversion of the arachidonic acid to the epoxide LTA_4.

LTA_4 is an unstable intermediate that may undergo three different pathways: hydrolysis, conjugation with glutathione or generate other bioactive eicosanoids (Figure 2.5).

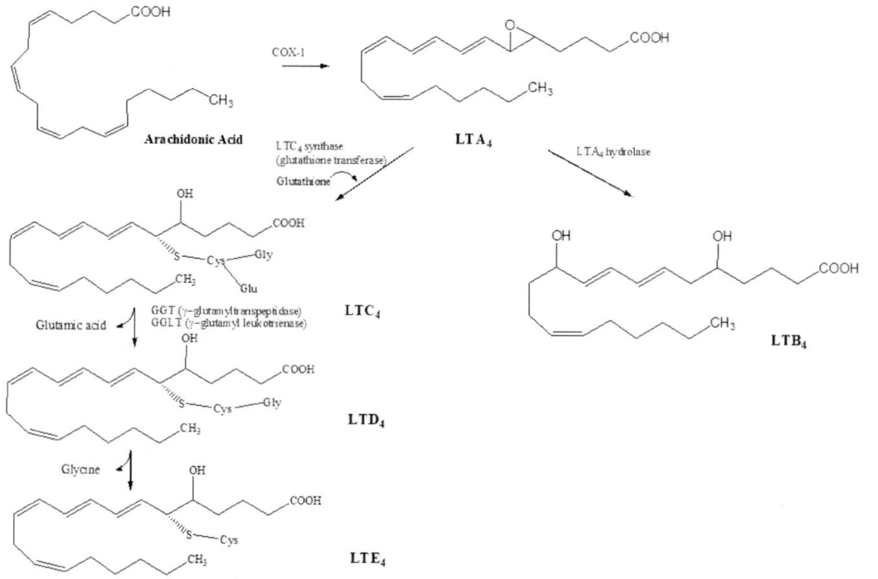

Figure 2.5. Pathway of leukotrienes synthesis.

Hydrolysis is catalysed by the enzyme leukotriene A_4 hydrolase in the cytoplasm, in order to get LTB_4. This eicosanoid is taken out from cells by a transporter. LTB_4 can act on neutrophils through receptors or stimulate leukocyte adhesion to endothelial cells. LTB_4 also acts intracellularly as it links to the nuclear transcription factor PPAR to induce transcription of target genes involved in β-oxidation of fatty acids, resulting LTB_4 degradation, as a homeostatic control of its pro-inflammatory actions.

Conjugation with glutathione renders LTC_4 and is catalysed by the enzyme LTC_4 synthase. LTC_4 is taken out of cells by specific transporters and suffers further extracellular metabolism by transpeptidases, forming LTD_4 and LTE_4 (Peters-Golden & Brock, 2003; Peters-Golden & Henderson, 2007). All of them are known as cysteinyl-leukotrienes (cys-LTs) and have effects on smooth muscle contraction. They have a potent effect on the constriction of bronchioles through cysteinyl leukotriene receptor 1, expressed in bronchiolar smooth muscle cells (Duroudier et al., 2009).

2.2.4. Functions of Prostanoids

As commented on before, the areas where the eicosanoids have demonstrated importance are: pain, fever and inflammation, acting mainly on the central nervous system (CNS), cardiovascular system, respiratory system, reproductive system or the kidneys.

One of the primary functions of the prostanoids is their relationship with the cardiovascular system. The role of prostanoids in cardiac dysfunction remains controversial; however, prostanoids appear to be therapeutically useful in various cardiovascular diseases such as arrhythmia, myocardial infarction, certain types of atherosclerosis and angina, among others. In turn, prostacyclin (PGI_2) and PGs, E_1, E_2 and D_2 are considered beneficial, playing an important role in inhibiting the progression of thrombus, promoting coronary vasodilation. On the other hand, prostanoids such as PGF_{2a} and TXA_2 are considered harmful for their vasoconstrictor activity; therefore, their inhibition may be beneficial [21, 34]. The primary PGs have a short halflife, being metabolised and excreted rapidly. Thus the PGFM tetranor can be quantified in urine as the mayor metabolite of PGF_2, which exerts vasoconstricting effects in bronchi, uterus, gastrointestinal tract and veins, among others [33].

In this regard, certain prostanoids act on vascular smooth muscle. In fact, PGE_2 and PGI_2 are vasodilator molecules while $PGF_{2\alpha}$ and TXA_2 produce vasoconstriction, acting mainly in vascular smooth muscles. Similarly, the PGs affect the respiratory system, relaxing the smooth muscle of the airways by the action of PGE_2 and PGI_2. In contrast, the airways are contracted by the action of TXA_2 and $PGF_{2\alpha}$ [35]. So 11-dehydro-TXB_2, derived metabolite of TXA_2, produces platelet release and potent vasoconstrictor (Catella et al., 1986). In this regard, PGEM tetranor is the metabolite synthesis of PGE2 and PGE1, PGs with vasodilator actions, activating and modulating inflammatory immune response [18]. Similarly, PGI_2 synthesized in the vascular endothelium and renal cortex, has vasodilator and antiplatelet very similar to those functions of PGD_2 [33].

Another one of the key aspects of prostanoids is their action on platelet aggregation, conditioned by the concentration of PGs. PGI_2 exerting antiplatelet and vasodilatory effects, while TXA_2 is a potent vasoconstrictor and platelet aggregator (Demers, 1984).

The effects of PGs in the central nervous system (CNS) are still poorly understood. However, some studies have shown the presence of prostanoids in certain diseases related to CNS. PGE_2 is the most common PG in the CNS, being produced at the neuronal level and plays an important role in the generation of fever, hyperalgesia and aids in the acquisition of memory [15]. Furthermore, it is related to the course of neurodegenerative diseases such as Parkinson disease and Alzheimer disease [36, 37].

Finally, certain PGs have also showed functions in the reproductive system, specifically in the womb. Processes such as menstruation, luteolysis or the onset of labor, are conditioned by the participation of $PGF_{2\alpha}$, and to a lesser extent by PGE_2 and PGI_2 [13]. The function of PGs has also been described in men. For example, PGE_2 plays an important role in the erection, ejaculation and transportation of sperm. Another point of interest is the renal system. Both PGE_2 and PGI_2 regulate renal blood flow, producing diuresis and natriuresis [12].

REFERENCES

[1] Voet, D, JG Voet, CW Pratt. 2006. "Lipids and biological membranes." *Fundamentals of Biochemistry. John Wiley and Sons, Inc., Hoboken, NJ, USA*:233-283.

[2] Von Euler, US. 1936. "On the specific vaso-dilating and plain muscle stimulating substances from accessory genital glands in man and certain animals (prostaglandin and vesiglandin)." *The Journal of physiology* no. 88 (2):213.

[3] Harizi, Hedi, Jean-Benoît Corcuff, Norbert Gualde. 2008. "Arachidonic-acid-derived eicosanoids: roles in biology and immunopathology." *Trends in molecular medicine* no. 14 (10):461-469.

[4] Strauss III, Jerome F, Robert L Barbieri. 2013. *Yen & Jaffe's Reproductive Endocrinology: Physiology, Pathophysiology, and Clinical Management*: Elsevier Health Sciences.

[5] Hernandez, Angel Gil DRT. 2010. *Tratado de nutrición: Nutrición clínica*. Vol. 4: Ed. Médica Panamericana.

[6] Miura, Yoshiro. 2013. "The biological significance of ω-oxidation of fatty acids." *Proceedings of the Japan Academy. Series B, Physical and biological sciences* no. 89 (8):370.

[7] Świzdor, Alina, Anna Panek, Natalia Milecka-Tronina, Teresa Kołek. 2012. "Biotransformations utilizing β-oxidation cycle reactions in the synthesis of natural compounds and medicines." *International journal of molecular sciences* no. 13 (12):16514-16543.

[8] Canosa, Elena Feduchi. 2014. *Bioquímica: conceptos esenciales*: Ed. Médica Panamericana.

[9] Wanders, Ronald JA. 2002. "Carnitine biosynthesis in mammals." *Biochemical Journal* no. 361 (3):417-429.

[10] Strijbis, Karin, Frédéric M Vaz, Ben Distel. 2010. "Enzymology of the carnitine biosynthesis pathway." *IUBMB life* no. 62 (5):357-362.

[11] Frungieri, Mónica B, Ricardo S Calandra, Artur Mayerhofer, María E Matzkin. 2015. "Cyclooxygenase and prostaglandins in somatic cell populations of the testis." *Reproduction* no. 149 (4):R169-R180.

[12] Spector, D., R. M. Zusman, B. V. Caldwell, L. Speroff. 1974. "The distribution of prostaglandins A, E, and F in the human kidney." *Prostaglandins* no. 6 (3):263-270.

[13] Sales, K. J., H. N. Jabbour. 2003. "Cyclooxygenase enzymes and prostaglandins in pathology of the endometrium." *Reproduction* no. 126 (5):559-567.

[14] Teismann, P., B. Ferger. 2001. "Inhibition of the cyclooxygenase isoenzymes COX-1 and COX-2 provide neuroprotection in the MPTP-mouse model of Parkinson's disease." *Synapse* no. 39 (2):167-174.

[15] Rall, J. M., S. A. Mach, P. K. Dash. 2003. "Intrahippocampal infusion of a cyclooxygenase-2 inhibitor attenuates memory acquisition in rats." *Brain Research* no. 968 (2):273-276.

[16] Bazan, N. G., V. Colangelo, W. J. Lukiw. 2002. "Prostaglandins and other lipid mediators in Alzheimer's disease." *Prostaglandins and Other Lipid Mediators* no. 68-69:197-210.

[17] Demers, L. M. 1984. "Prostaglandins in human disease." *Clinics in Laboratory Medicine* no. 4 (4):787-801.

[18] Honda, Hideo, Kazunaga Fukawa, Takashi Sawabe. 1980. "Influence of adjuvant arthritis on main urinary metabolites of prostaglandin F and E in rats." *Prostaglandins* no. 19 (2):259-269.

[19] Catella, F., D. Healy, J. A. Lawson, G. A. FitzGerald. 1986. "11-Dehydrothromboxane B2: A quantitative index of thromboxane A2 formation in the human circulation." *Proceedings of the National Academy of Sciences of the United States of America* no. 83 (16):5861-5865.

[20] Miller, S. B. 2006. "Prostaglandins in Health and Disease: An Overview." *Seminars in Arthritis and Rheumatism* no. 36 (1):37-49.

[21] Chan, Peter S, Peter Cervoni. 1986. "Prostaglandins, prostacyclin, and thromboxane in cardiovascular diseases." *Drug Development Research* no. 7 (4):341-359.

[22] Beamish, R. E., P. K. Das, M. Karmazyn, N. S. Dhalla. 1985. "Prostaglandins and heart disease." *Canadian Journal of Cardiology* no. 1 (1):66-74.

[23] Cuenca, E Martín. 2006. *Fundamentos de fisiología*: Editorial Paraninfo.

[24] Zhou, W., T. S. Blackwell, K. Goleniewska, J. F. O'Neal, G. A. FitzGerald, M. Lucitt, R. M. Breyer, R. S. Peebles Jr. 2007. "Prostaglandin I2 analogs inhibit Th1 and Th2 effector cytokine production by CD4 T cells." *Journal of Leukocyte Biology* no. 81 (3):809-817.

[25] Evans, J. H., D. M. Spencer, A. Zweifach, C. C. Leslie. 2001. "Intracellular Calcium Signals Regulating Cytosolic Phospholipase A 2 Translocation to Internal Membranes." *Journal of Biological Chemistry* no. 276 (32):30150-30160.

[26] Smith, W. L. 1989. "The eicosanoids and their biochemical mechanisms of action." *Biochemical Journal* no. 259 (2):315-324.

[27] Smith, W. L., D. L. DeWitt, R. M. Garavito. 2000. Cyclooxygenases: Structural, cellular, and molecular biology. In *Annual Review of Biochemistry*.

[28] Smith, W. L., Y. Urade, P. J. Jakobsson. 2011. "Enzymes of the cyclooxygenase pathways of prostanoid biosynthesis." *Chemical Reviews* no. 111 (10):5821-5865.

[29] Gómez-Luque, A. 2005. "Inhibidores de la COX¿ hacia dónde vamos?" *Revista de la Sociedad Española del Dolor* no. 12 (6):321-325.

[30] Chandrasekharan, NV, Daniel L Simmons. 2004. "The cyclooxygenases." *Genome biology* no. 5:/2004/5/9/241-/2004/5/9/241.

[31] Van der Donk, Wilfred A, Ah-Lim Tsai, Richard J Kulmacz. 2002. "The cyclooxygenase reaction mechanism." *Biochemistry* no. 41 (52):15451-15458.

[32] Rouzer, Carol A, Lawrence J Marnett. 2003. "Mechanism of free radical oxygenation of polyunsaturated fatty acids by cyclooxygenases." *Chemical Reviews* no. 103 (6):2239-2304.

[33] Funk, Colin D. 2001. "Prostaglandins and leukotrienes: advances in eicosanoid biology." *Science* no. 294 (5548):1871-1875.

[34] Beamish, RE, PK Das, M Karmazyn, NS Dhalla. 1985. "Prostaglandins and heart disease." *The Canadian journal of cardiology* no. 1 (1):66-74.

[35] Miller, Stephen B. 2006. Prostaglandins in health and disease: an overview. Paper read at Seminars in arthritis and rheumatism.

[36] Teismann, Peter, Boris Ferger. 2001. "Inhibition of the cyclooxygenase isoenzymes COX-1 and COX-2 provide neuroprotection in the MPTP-mouse model of Parkinson's disease." *Synapse* no. 39 (2):167-174.

[37] Bazan, Nicolas G, Vittorio Colangelo, Walter J Lukiw. 2002. "Prostaglandins and other lipid mediators in Alzheimer's disease." *Prostaglandins & other lipid mediators* no. 68:197-210.

Chapter 3

FORMATION OF LIPID PEROXIDATION MARKERS

Javier Marhuenda Hernández, PhD

ABSTRACT

An increase in prooxidant species could lead to membrane damage due to its atack on polyunsaturated fatty acids, which are easily oxidizable owing to their methylated interrupted double bonds, that can participate in the formation of acyclic, and subsequently, stable cyclic oxygenated metabolites. Arachidonic acid, adrenic acid, eicosapentaenoic acid and docosahexanoic acid are commonly-studied fatty acids. The oxidative attack to those fatty acids results in F_2-isoprostanes, F_2-dihomo-isoprostanes, F_3-neuroprostanes and F_4-neuroprostanes respectively.

Nowadays, elevated levels of F_2-isoprostanes in biological fluids is recognized as the reference biomarker for lipid peroxidation and oxidative stress measurement. Not only F_2-isoprostanes, but also neuroprostanes and F_2-dihomo-isoprostanes are useful biomarkers for the measurement of oxidative injury.

Arachidonic acid is widely distirbuted along the whole organism, while adrenic acid, eicosapentaenoic acid and docosahexanoic acid are mainly found in the central nervious system. Therefore, while the mesurement of F_2-isoprostanes allows one to know the oxidative damage produced at a sistemic level, neuroprostanes and F_2-dihomo-isoprostanes are representative of an oxidative atack to the central nervious system.

Keywords: polyunsaturated fatty acids, oxidative attack, isoprostanes, F_2-dihomo-isoprostanes, neuroprostanes

INTRODUCTION

Reactive oxygen species (ROS) and/or nitrogen reactive species (NOS) lead to oxidative reactions, causing oxidative stress in the organism [1]. Under such conditions, many molecules and structures are susceptible to oxidation, causing dysfunctions and certain disorders in the organism. Among lipids, polyunsaturated fatty acids (PUFAs) have methylated interrupted double bonds that can participate to form acyclic, and subsequently, stable cyclic oxygenated metabolites [3]. Therefore, central nervous system structures like neurons are also oxidizable due to their abundant content of PUFAs. An increase in ROS/NOS could cause neural membrane damage, leading to the development of different neuropathologies - including Alzheimer's disease (AD), Parkinson's disease (PD), or multiple sclerosis (MS) [4].

Lipid peroxidation products have been investigated in order to determine their use as biomarkers of the oxidative status within the human body. Linolenic acid (LNA), arachidonic acid (AA), adrenic acid (AdA), eicosapentaenoic acid (EPA) and docosahexanoic acid (DHA) are commonly-studied fatty acids [3, 5, 6] which can give rise to phytoprostanes (PhytoPs) [5, 7], F_2-isoprostanes (F_2-IsoPs) [3, 6], F_2-dihomo-isoprostanes (F_2-dihomo-IsoPs), F_3-neuroprostanes (F_3-NeuroPs) and F_4-neuroprostanes (F_4-NeuroPs) [3], respectively.

There are many biomarkers for the measurement of oxidative injury on an organism such as thiobarbituric acid or malondialdehyde. However, an elevation of F_2-IsoP levels in biological fluids is accepted as the reference biomarker for lipid peroxidation and oxidative stress [6]. Due to reactivity and short lifespan of free radicals, oxidative stress must be assessed by the measurement of damaged biological products, considered as biomarkers of lipid peroxidation.

3.1. SYNTHESIS OF ISOPROSTANES

The first research of Isoprostanes (IsoPs) took place in 1975 by Porter and Funk, other than Pryor and Stanley during their research about the biosynthesis *in vitro* of prostaglandins [8, 9]. They named those new molecules PG-like compounds. The first *in vivo* assay in humans was reported ten years later, Liston et al. (1985) showed that enzymatic oxidation of PGD_2 by COX mainly produces $9\alpha,11\beta\text{-}PGF_{2\alpha}$ by 11-ketoreductase. Nevertheless, Wendelborn et al. (1988) reported the instability of PGD_2 under aqueous conditions, leading to isomers of $9\alpha,11\beta\text{-}PGF_{2\alpha}$, previous isomerization and their final reduction. The analysis of plasma samples by mass spectrometry carried out by Morrow et al. (1990) revealed a series of peaks owning characteristics of F-ring PGs. They reported an increment up to 100-fold higher on the areas of these peaks when samples were frozen at -20 °C during few months. Moreover, a study in rats reported that antioxidants and reducing agents showed a reduction in the initial concentration of these PGF_2-like compounds; while at the same time, a toxic compound (CCl_4) producing free radical was able to increase their formation [12, 13]. These previous studies revealed the *in vivo* formation of F_2-IsoPs by non-enzymatically peroxidation of arachidonic acid, despite its enzymatic oxidation by COX.

The term IsoPs was entered by Morrow et al. (1990), suggesting a non-enzymatic oxidative pathway of arachidonic acid, dependent on phospholipids [12, 14]. The F_2-IsoPs series are the most studied IsoPs, formed by phospholipases as acetylhydrolase, being released as free acids [15, 16] which are transported by high density lipoproteins [17] in plasma. Finally, they are excreted in urine mainly as glucuronides [18].

Morrow et al. (1994) reported a mechanism explaining the pathway of the F_2-IsoPs from peroxidation of arachidonic acid (Figure 1). The first step consists of the formation of the peroxyl radical, due to the addition of molecular oxygen and previous abstraction of a bisallylic hydrogen atom. In the second step, the peroxyl radical suffers 5-exo cyclization and a second addition of other molecular oxygen leads to the formation of unstable PGG_2-like compounds, which are reduced to F_2-IsoPs. In theory, there are four F_2-IsoPs regioisomers (5-, 8-, 12- and 15- series), which lead to 8 racemic diastereoisomers, producing 64 possible compounds (Figure 1). Both *in vivo* and *in vitro* studies showed higher content of the 5- and 15- series in contrast to the 8- and 12- series [20, 21], owing to the rapid metabolization of the 8- and 12- series IsoPs by human organism [22]. This nomenclature classification has been accepted by the Eicosanoid Nomenclature Committee, certified by JCBN of IUPAC [23]. However, Rokach et al. (1997) suggested other nomenclature based on the abbreviation iP for identifying isoprostane, designating the regioisomers as III-VI (Figure 3.1).

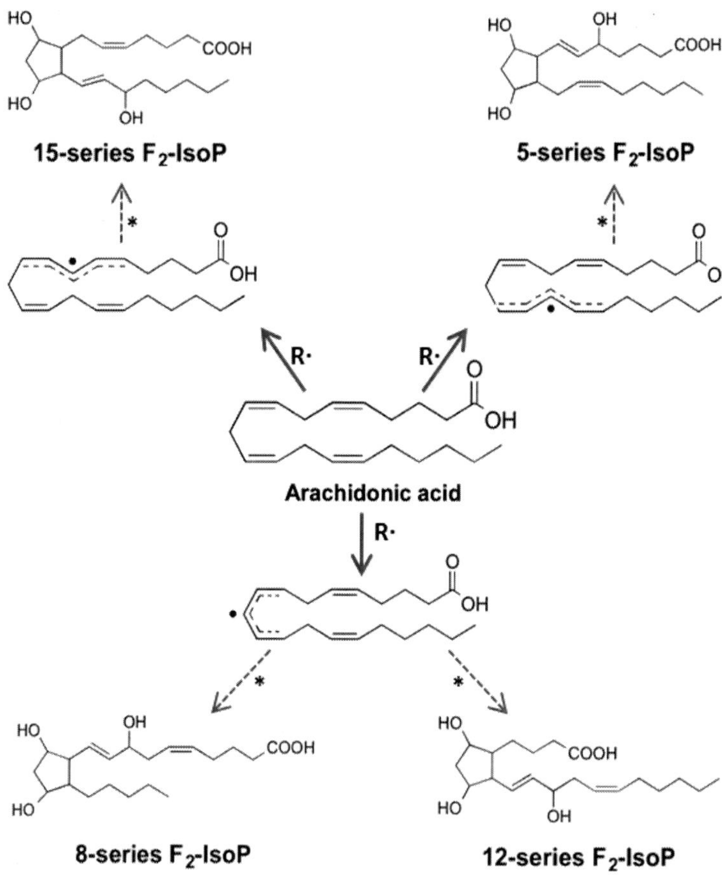

Figure 3.1. Synthesis of F_2-Isoprostanes from the peroxidation of arachidonic acid. *Reactions that occurred are: cyclization, oxygen addition and reduction to F_2-IsoPs. Adapted from Milne et al. (2014).

Not only F_2-IsoPs but also other IsoPs have been described, differing on the functional groups on the prostane ring (Figure 3.2). Two other IsoPs are formed from the isomerization of the arachidonyl endoperoxide 1, named E_2- and D_2-IsoPs, isomeric forms of PGE_2 and PGD_2 [25]. Moreover, while F_2-IsoPs are formed under normal physiological conditions, E_2-IsoPs and D_2-IsoPs came from Kornblum–DeLaMare readjustment [26]. The diminushing of glutathione (GSH) or α-tocopherol -cellular reducing mediators- leads to the synthesis of E_2-IsoPs and D_2-IsoPs. Meanwhile, F_2-IsoPs are formed via reduction of the endoperoxide 1 [27]. Contrary to F_2-IsoPs, E_2-IsoPs and D_2-IsoPs, they are not terminal products of the IsoPs pathway. Dehydration of E_2-

IsoPs and D_2-IsoPs is easily produced under physiological conditions, and in fact that leads to the formation of cyclopentenones A_2-IsoPs and J_2-IsoPs respectively, both in vitro and in vivo [28, 29]. Finally, A_2-IsoPs and J_2-IsoPs are further rearranged and dehydrated, leading to deoxy-A_2-IsoPs and deoxy-J_2-IsoPs [30].

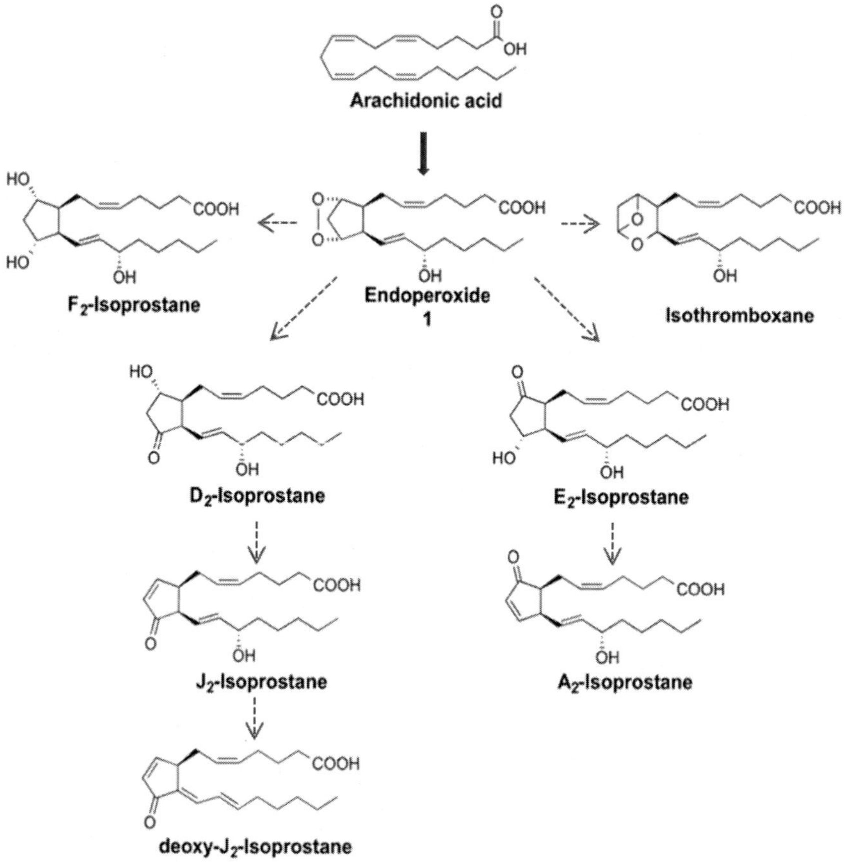

Figure 3.2. Synthesis of the families of isoprostanes from the peroxidation of arachidonic acid. Adapted from Milne et al. (2014).

In addition to IsoPs, other compounds are formed from non-enzymatic oxidation. Isothromboxanes are isomeric compounds to thromboxane B_2 (catalyzed by COX), but are generated by the free radical-catalyzed peroxidation of arachidonic acid [31]. An additional group of related compounds are also produced from the peroxidation of arachidonic acid, named isofurans (IsoFs). While IsoPs contain a cyclopentane ring formed by 5-exo-cyclization of the carbon-centered radical, IsoFs hold a substituted tetrahydrofuran ring, also formed by the 5-exo-cyclization of the carbon-centered radical and favoured by increased oxygen tension. The family of IsoFs is comprised of a total of eight regioisomers and its subsequent 16 racemic diastereomers, leading to a total of 256 compounds [32].

3.2. Synthesis of Neuroprostanes and F_2-Dihomo-Isoprostanes

Since the discovery of IsoPs in 1975, the scientific community has reported some other peroxidation products derived from the attack of free radicals to many other PUFAs. ALA, EPA, AdA and DHA have at least two-skipped diene units, being able to form isoprostane-like compounds [33]. The attack of free radicals on DHA and EPA, the main Ω-3 PUFAs present in humans, produce F_4- and F_3-NeuroPs [34, 35]. Other Ω-6 PUFA, mainly present in brain white matter, is AdA, which produces dihomo-IsoPs when subjected to radical attack [36].

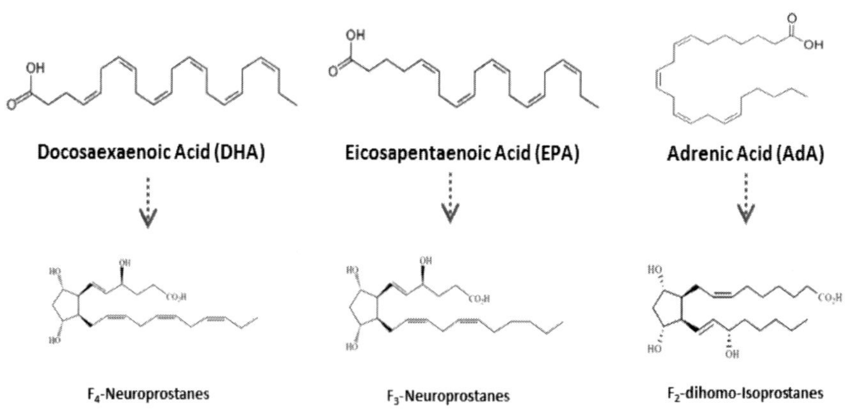

Figure 3.3. Formation of neuroprostanes and F_2-dihomo-isoprostanes.

Similar to IsoPs, the formation of NeuroPs and F_2-dihomo-IsoPs leads to a G- type intermediate metabolite. In addition to the F- type, the peroxidation of DHE, EPA and AdA could lead to, E-, D-, A-, B-, and L- NeuroPs and F_2-dihomo-IsoPs depending on the close conditions of the membrane phospholipids (Figure 3.3) [33]. As commented on above, Fessel et al. (2002) revealed for the first time that the formation of IsoFs from ARA are due to an increment of molecular oxygen after an initial cyclization of the peroxidation process. In the same way, such an increment of the oxygen concentration during the peroxidation of DHA and AdA could lead to the formation of neurofurans (NeuroFs) [37, 38] and dihomo-Isofurans (dihomo-IsoFs) [39] respectively.

Therefore, NeuroPs and F_2-dihomo-IsoPs offer a distinctive field for the measurement and the quantification of oxidative stress injury that's accomplished mainly in neuronal membranes in vivo. This approach is profoundly significant when considering neurodegenerative diseases and its development in the central nervous system.

REFERENCES

[1] Halliwell B, Gutteridge JMC. 2007. *Free radicals in biology and medicine*. 4th ed: ed. Oxford: Oxford university press.

[2] Milne, Ginger L., Qi Dai, L. Jackson Roberts Ii. 2014. "*The isoprostanes—25 years later." Biochimica et Biophysica Acta (BBA) - Molecular and Cell Biology of Lipids* (0). doi: http://dx.doi.org/10.1016/j.bbalip.2014.10.007.

[3] Miller, E., A. Morel, L. Saso, J. Saluk. 2014. "*Isoprostanes and neuroprostanes as biomarkers of oxidative stress in neurodegenerative diseases.*" Oxidative Medicine and Cellular Longevity no. 2014.

[4] Farooqui, T., A. A. Farooqui. 2011. "Lipid-mediated oxidative stress and inflammation in the pathogenesis of Parkinson's disease." *Parkinson's Disease*. doi: 10.4061/2011/247467.

[5] Imbusch, R., M. J. Mueller. 2000. "Formation of isoprostane F2-like compounds (phytoprostanes F1) from α-linolenic acid in plants." *Free Radical Biology and Medicine* no. 28 (5):720-726.

[6] Medina, S., R. Domínguez-Perles, J. I. Gil, F. Ferreres, C. García-Viguera, J. M. Martínez-Sanz, A. Gil-Izquierdo. 2012. "A ultra-pressure liquid chromatography/triple quadrupole tandem mass spectrometry method for the analysis of 13 eicosanoids in human urine and

quantitative 24 hour values in healthy volunteers in a controlled constant diet." *Rapid Communications in Mass Spectrometry* no. 26 (10):1249-1257.

[7] Marhuenda, Javier, Sonia Medina, Alexandra Díaz-Castro, Pedro Martínez-Hernández, Simón Arina, Pilar Zafrilla, Juana Mulero, Camille Oger, Jean-Marie Galano, Thierry Durand. 2015. "Dependency of Phytoprostane Fingerprints of Must and Wine on Viticulture and Enological Processes." *Journal of Agricultural and Food Chemistry* no. 63 (41):9022-9028.

[8] Porter, NA, MO Funk. 1975. "Letter: Peroxy radical cyclization as a model for prostaglandin biosynthesis." *The Journal of organic chemistry* no. 40 (24):3614.

[9] Pryor, William A, JP Stanley. 1975. "Suggested mechanism for the production of malonaldehyde during the autoxidation of polyunsaturated fatty acids. Nonenzymic production of prostaglandin endoperoxides during autoxidation." *The Journal of organic chemistry* no. 40 (24):3615-3617.

[10] Liston, T. E., J. A. Oates, L. J. Roberts 2nd. 1985. "Prostaglandin D2 is metabolized in humans to 9 alpha,11 beta-prostaglandin F2, a novel biologically active prostaglandin." *Advances in prostaglandin, thromboxane, and leukotriene research* no. 15:365-367.

[11] Wendelborn, D. F., K. Seibert, L. J. Roberts 2nd. 1988. "Isomeric prostaglandin F2 compounds arising from prostaglandin D2: a family of icosanoids produced in vivo in humans." Proceedings of the National Academy of Sciences of the United States of America no. 85 (2):304-308.

[12] Morrow, Jason D, Kristina E Hill, Raymond F Burk, Tarek M Nammour, Kamal F Badr, LJ 2nd Roberts. 1990. "A series of prostaglandin F2-like compounds are produced in vivo in humans by a non-cyclooxygenase, free radical-catalyzed mechanism." Proceedings of the National Academy of Sciences no. 87 (23):9383-9387.

[13] Morrow, J. D., J. A. Awad, T. Kato, K. Takahashi, K. F. Badr, L. Jackson Roberts Ii, R. F. Burk. 1992. "Formation of novel non-cyclooxygenase-derived prostanoids (F2-lsoprostanes) in carbon tetrachloride hepatotoxicity: An animal model of lipid peroxidation." *Journal of Clinical Investigation* no. 90 (6):2502-2507.

[14] Kadiiska, M. B., B. C. Gladen, D. D. Baird, D. Germolec, L. B. Graham, C. E. Parker, A. Nyska, J. T. Wachsman, B. N. Ames, S. Basu, N. Brot, G. A. FitzGerald, R. A. Floyd, M. George, J. W. Heinecke, G.

E. Hatch, K. Hensley, J. A. Lawson, L. J. Marnett, J. D. Morrow, D. M. Murray, J. Plastaras, L. J. Roberts Ii, J. Rokach, M. K. Shigenaga, R. S. Sohal, J. Sun, R. R. Tice, D. H. Van Thiel, D. Wellner, P. B. Walter, K. B. Tomer, R. P. Mason, J. C. Barrett. 2005. "Biomarkers of Oxidative Stress Study II: Are oxidation products of lipids, proteins, and DNA markers of CCl4 poisoning?" *Free Radical Biology and Medicine* no. 38 (6):698-710. doi: http://dx.doi.org/10.1016/j.freeradbiomed.2004.09.017.

[15] Stafforini, Diana M, James R Sheller, Timothy S Blackwell, Adam Sapirstein, Fiona E Yull, Thomas M McIntyre, Joseph V Bonventre, Stephen M Prescott, L Jackson Roberts. 2006. "Release of free F2-isoprostanes from esterified phospholipids is catalyzed by intracellular and plasma platelet-activating factor acetylhydrolases." *Journal of Biological Chemistry* no. 281 (8):4616-4623.

[16] Kono, Nozomu, Takao Inoue, Yasukazu Yoshida, Hiroyuki Sato, Tomokazu Matsusue, Hiroyuki Itabe, Etsuo Niki, Junken Aoki, Hiroyuki Arai. 2008. "Protection against oxidative stress-induced hepatic injury by intracellular type II platelet-activating factor acetylhydrolase by metabolism of oxidized phospholipids in vivo." *Journal of Biological Chemistry* no. 283 (3):1628-1636.

[17] Proudfoot, Julie M, Anne E Barden, Wai Mun Loke, Kevin D Croft, Ian B Puddey, Trevor A Mori. 2009. "HDL is the major lipoprotein carrier of plasma F2-isoprostanes." *Journal of lipid research* no. 50 (4):716-722.

[18] Yan, Zhao, Emilie Mas, Trevor A Mori, Kevin D Croft, Anne E Barden. 2010. "A significant proportion of F 2-isoprostanes in human urine are excreted as glucuronide conjugates." *Analytical biochemistry* no. 403 (1):126-128.

[19] Morrow, J. D., T. A. Minton, C. R. Mukundan, M. D. Campbell, W. E. Zackert, V. C. Daniel, K. F. Badr, I. A. Blair, L. J. Roberts Ii. 1994. *Free radical-induced generation of isoprostanes in vivo: evidence for the formation of D-ring and E-ring isoprostanes.*

[20] Waugh, R. J., J. D. Morrow, L. J. Roberts Ii, R. C. Murphy. 1997. "Identification and relative quantitation of F2-isoprostane regioisomers formed in vivo in the rat." Free Radical Biology and Medicine no. 23 (6):943-954. doi: 10.1016/S0891-5849(97)00133-0.

[21] Li, H., J. A. Lawson, M. Reilly, M. Adiyaman, S. W. Hwang, J. Rokach, G. A. FitzGerald. 1999. "Quantitative high performance liquid chromatography/tandem mass spectrometric analysis of the four classes

of F2-isoprostanes in human urine." Proceedings of the National Academy of Sciences of the United States of America no. 96 (23):13381-13386. doi: 10.1073/pnas.96.23.13381.

[22] Yin, H., J. D. Morrow, N. A. Porter. 2004. "Identification of a Novel Class of Endoperoxides from Arachidonate Autoxidation." *Journal of Biological Chemistry* no. 279 (5):3766-3776. doi: 10.1074/jbc.M307137200.

[23] Taber, D. F., J. D. Morrow, L. J. Roberts Ii. 1997. "A nomenclature system for the isoprostanes." *Prostaglandins* no. 53 (2):63-67. doi: 10.1016/S0090-6980(97)00005-1.

[24] Rokach, J., S. P. Khanapure, S. W. Hwang, M. Adiyaman, J. A. Lawson, G. A. Fitzgerald. 1997. "Nomenclature of isoprostanes: A proposal." *Prostaglandins* no. 54 (6):853-873. doi: 10.1016/S0090-6980(97)00184-6.

[25] Morrow, J. D., T. A. Minton, C. R. Mukundan, M. D. Campbell, W. E. Zackert, V. C. Daniel, K. F. Badr, I. A. Blair, L. J. Roberts Ii. 1994. "Free radical-induced generation of isoprostanes in vivo: evidence for the formation of D-ring and E-ring isoprostanes." *Journal of Biological Chemistry* no. 269 (6):4317-4326.

[26] Kornblum, N., H. E. DeLaMare. 1951. "The base catalyzed decomposition of a dialkyl peroxide [7]." *Journal of the American Chemical Society* no. 73 (2):880-881.

[27] Morrow, J. D., L. J. Roberts, V. C. Daniel, J. A. Awad, O. Mirochnitchenko, L. L. Swift, R. F. Burk. 1998. "Comparison of formation of D2/E2-isoprostanes and F2-isoprostanes in vitro and in vivoe -- Effects of oxygen tension and glutathione." *Archives of Biochemistry and Biophysics* no. 353 (1):160-171. doi: 10.1006/abbi.1998.0645.

[28] Chen, Y., W. E. Zackert, L. J. Roberts Ii, J. D. Morrow. 1999. "Evidence for the formation of a novel cyclopentenone isoprostane, 15-A(2t)-isoprostane (8-iso-prostaglandin A2) in vivo." *Biochimica et Biophysica Acta - Molecular and Cell Biology of Lipids* no. 1436 (3):550-556. doi: 10.1016/S0005-2760(98)00168-4.

[29] Chen, Y., J. D. Morrow, L. J. Roberts Ii. 1999. "Formation of reactive cyclopentenone compounds in vivo as products of the isoprostane pathway." *Journal of Biological Chemistry* no. 274 (16):10863-10868. doi: 10.1074/jbc.274.16.10863.

[30] Hardy, K. D., B. E. Cox, G. L. Milne, H. Yin, L. J. Roberts Ii. 2011. "Nonenzymatic free radical-catalyzed generation of 15-deoxy-Δ 12,14-Prostaglandin J 2-like compounds (deoxy-J 2-isoprostanes) in vivo 1." *Journal of lipid research* no. 52 (1):113-124. doi: 10.1194/jlr.M010264.

[31] Morrow, J. D., J. A. Awad, A. Wu, W. E. Zackert, V. C. Daniel, L. Jackson Roberts Ii. 1996. "Nonenzymatic free radical-catalyzed generation of thromboxane-like compounds (Isothromboxanes) in vivo." *Journal of Biological Chemistry* no. 271 (38):23185-23190. doi: 10.1074/jbc.271.38.23185.

[32] Fessel, J. P., N. A. Porter, K. P. Moore, J. R. Sheller, L. J. Roberts Ii. 2002. "Discovery of lipid peroxidation products formed in vivo with a substituted tetrahydrofuran ring (isofurans) that are favored by increased oxygen tension." Proceedings of the National Academy of Sciences of the United States of America no. 99 (26):16713-16718. doi: 10.1073/pnas.252649099.

[33] Jahn, U., J. M. Galano, T. Durand. 2008. "Beyond prostaglandins - Chemistry and biology of cyclic oxygenated metabolites formed by free-radical pathways from polyunsaturated fatty acids." *Angewandte Chemie - International Edition* no. 47 (32):5894-5955. doi: 10.1002/anie.200705122.

[34] Nourooz-Zadeh, J., E. H. C. Liu, E. E. Änggård, B. Halliwell. 1998. "F4-isoprostanes: A novel class of prostanoids formed during peroxidation of docosahexaenoic acid (DHA)." *Biochemical and Biophysical Research Communications* no. 242 (2):338-344. doi: 10.1006/bbrc.1997.7883.

[35] Roberts, L Jackson, Thomas J Montine, William R Markesbery, Andrew R Tapper, Pierre Hardy, Sylvain Chemtob, Wolff D Dettbarn, Jason D Morrow. 1998. "Formation of isoprostane-like compounds (neuroprostanes) in vivo from docosahexaenoic acid." *Journal of Biological Chemistry* no. 273 (22):13605-13612.

[36] Vanrollins, M., R. L. Woltjer, H. Yin, J. D. Morrow, T. J. Montine. 2008. "F 2-Dihomo-isoprostanes arise from free radical attack on adrenic acid." *Journal of lipid research* no. 49 (5):995-1005. doi: 10.1194/jlr.M700503-JLR200.

[37] Arneson, Kyle O, L Jackson Roberts. 2007. "Measurement of products of docosahexaenoic acid peroxidation, neuroprostanes, and neurofurans." *Methods in Enzymology* no. 433:127-143.

[38] Song, Wen-Liang, John A Lawson, Dermot Reilly, Joshua Rokach, Chih-Tsung Chang, Benoit Giasson, Garret A FitzGerald. 2008. "Neurofurans, novel indices of oxidant stress derived from docosahexaenoic acid." *Journal of Biological Chemistry* no. 283 (1):6-16.

[39] de La Torre, Aurélien, Yiu Yiu Lee, Camille Oger, Per Torp Sangild, Thierry Durand, Jetty Chung-Yung Lee, Jean-Marie Galano. 2014. "Synthesis, Discovery, and Quantitation of Dihomo-Isofurans: Biomarkers for In Vivo Adrenic Acid Peroxidation." *Angewandte Chemie* no. 126 (24):6363-6366.

Chapter 4

LIPID PEROXIDATION PRODUCTS AS BIOMARKERS IN DIFFERENT PATHOLOGIES

Javier Marhuenda Hernández, PhD

ABSTRACT

The peroxidation products of polyunsaturated fatty acids are remarkable biomarkers of oxidative stress. Due to the diverse distribution and location of the different polyunsaturated fatty acids, the measurement of a specific peroxidation product of each one, could lead to a different comprehension of the location of the oxidative injury.

The scientific community has developed several studies regarding the urinary excretion of isoprostanes and their concentration in plasma and cerebrospinal fluid. Many neurodegenerative diseases such as Alzheimer disease, Parkinson disease, cerebrovascular injury, Huntington disease or epilepsy have been shown to be determined by elevated levels of proxidation markers. In addition, this is seen in other diseases like obesity or cancer, besides genetical or dietetical disorders.

Finally, the role of the different peroxidation products will also discussed. Although limited, the knowledge about the biological role of isoprostanes is quite realted to what's observed for diverse proinflammatory prostaglandines. However, the knowledge about neuroprostanes and dihomo-isoprostanes is still limited.

Keywords: Parkinson disease, Alzheimer disease, biological function, F_2-isoprostanes, neuroprostanes, F_2-dihomo-isoprostanes

Introduction

Biomarkers are defined as the reliable indicators of normal biological or pathologic processes, which can be easily and objectively measured and evaluated [1]. The utilization of biomarkers is essential on the detection of diseases and the assessment of treatment response. One important factor for choosing a biomarker is non invasive extraction, in order for the measurement of circulating levels of specific biomarkers in a wide range of diseases.

Among other functions, the biomarkers might be used for precise measurement of oxidative stress *in vivo* [2]. Of all the different molecules that make up the tissues of the human organism, polyunsaturated fatty acids seem to be the most vulnerable to the attack of oxidant species [3, 4].

The attack of lipids by free radicals causes unsaturated fatty acid to react directly with molecular oxygen [3]. This process is named lipid peroxidation, being monitored by the insertion of an oxygen molecule, which acts as fuel to free radicals. This type of lipid oxidation has been recognised since antiquity as a problem in the storage of fats and oils, and peroxidation products of lipids has long been studied within scientific literature, comprising of many fields (food chemistry, polymer chemistry and even restorers studing the deterioration of valuable paintings).

Consequently, the formation of lipid peroxidation molecules can be used as a biomarker for the measurement of oxidative stress status in vivo [5]. The levels of lipid peroxidation products such as isoprostanes, neuroprostanes and dihomo-isoprostanes from biological human fluids/tissues have been extensively reported by the scientific community [6].

Currently, different lipid peroxidation products are used for assessment of oxidative stress *in vivo* and, depending on the disease, a specific peroxidation product could be more adequate than others.

4.1. Oxidative Stress Biomarkers

Traditionally, the measurement of oxidative injuries has been made difficult due to the lack of reliable biomarkers which were ubiquitous in the organism and chemically stable in biological fluids under correct storage environments. However, since the discovery of IsoPs, they have been recognized as the "gold standard" for measuring lipid peroxidation in particular and oxidative stress in general.

The consistency of IsoPs as oxidative stress biomarkers has been verified both in animal and human studies. Kadiiska et al. (2005) found a dose-dependent increase of IsoPs after the administration of carbon tetrachloride in mice. In turn, Il'Yasova et al. (2010) reported increased urinary excretion rate of total F_2-IsoPs and 2,3-dinor-15-F_{2t}-IsoP in 23 cancer-diagnosed women. The levels in urine increased 1 hour after treatment with standard chemotherapy and returned to their initial values after 24 hours.

The production of IsoPs can be measured in multiple biological fluids, including: plasma, urine, cerebrospinal fluid, amniotic fluid, saliva and exhaled breath condensate [6]. However, the most suitable fluids are plasma and urine since they are a slightly invasive methodology for its analysis. Moreover, the collection and storage of the samples is an important factor. The collection of the samples must be done while avoiding external factors that could lead to *ex vivo* oxidation. Finally, the storage of the samples should be done at -80 ºC, thus avoiding autoxidation of fatty acids which is not excluded at -20 ºC [9].

4.2. NEURODEGENERATIVE DISORDERS

The central nervous system (CNS) can be easily oxidized because of its high oxygen demand and the abundance of PUFAs, besides the minor abundance of endogenous antioxidant defenses in comparison with other tisues [10]. Neurodegenerative disorders are characterized by chronic inflammation and the subsequent accumulation of proinflammatory cytokines, stimulating an increase in the oxidative stress and neuronal damage [11]. Similar to CNS, the composition of fatty acids of neurons are mainly PUFAs. That fact made neurons easily oxidizable and susceptible to neural damage induced by free radicals. It has been previously reported that inflammatory processes could contribute to the increase of many neurodegenerative disorders such as Alzheimer's disease, Parkinson's disease, and multiple sclerosis [12-14]. The main intermediaries of inflammatory processes on the SNC are microglial cells, which are transformed in activated microglia in presence of oxidant compounds. An increment on the activation of microglia could turn into a chronic inflammation, due to the release of proinflammatory cytokines, free radicals, proteinases, PGE_2, etc [10]…

DHA, EPA and AdA are the main PUFAs of the CNS; therefore, the measurement of NeuroPs and dihomo-IsoPs is quite related to the specific damage on CNS. However, AA is widely distributed along all the structures in humans [15], so the measurement of IsoPs in plasma or urine does not reveal the specific site of formation of IsoPs. Therefore, it is necessary for the measurement of the cerebrospinal fluid (CSF) in order to obtain the correct identification of local oxidant injury, more than just oxidative stress' general effects [16, 17].

As commented on before, lipid peroxidation is a consequence of oxidative stress, being a characteristic for several diseases such as neurodegenerative disorders. In this section we will see the main neurodegenerative disorders associated with different biomarkers of lipid peroxidation.

4.2.1. Multiple Sclerosis

As commented above, MS is a heterogeneous disease that is developed by several factors and can be determined by various disorders such as: inflammation, demyelination, redox unpairrment, axonal damage or unvalid repair processes. The most common is relapsing-remitting course (RRMS), defined by inflammatory attacks on myelin as well as the nerve fibers themselves, in which relapse occurs from time to time followed up by complete or partial recovery [10]. That disorder comprises of over 80% of all the cases of MS and is characterized by general inflammation and edemas, owing to the liberation of proinflammatory cytoquines. RRMS progression derives from permanent neuropathological disorders after 10-20 years in a half of cases [18]. Therefore, that kind of MS is also known as secondary progressive multiple sclerosis (SPMS) [19]. On the contrary, a primary progressive multiple sclerosis (PPMS) patient suffers the progressive symtoms of the disease from its beginning. The deterioration of the myelin from axons is the primary basis of neurodegeneration in MS patients, leading to neuronal incapacity, such that neuronal incapacity is permanent even in patients with no trace of brain damage after a magnetic resonance scan. Therefore, oxidative injury may be the cause of this type of MS [2, 20].

As mentioned by Miller et al. (2014), the biomarkers of MS could be divided in seven categories: alteration of the immune system, axonal/neuronal damage, blood-brain barrier disruption, demyelination, gliosis, repaired remyelination and oxidative stress. Therefore, IsoPs could be good biomarkers for the monitorization of oxidative stress developed-MS. Many studies have shown an increase of IsoPs in MS patients. Santos et al. (2013) reported 6-fold higher urinary excretion of IsoPs in MS patients compared to control patients. Mattsson et al. (2007) measured the levels of 15-F_{2t}-IsoPs in CSF, reporting increased levels of MS patients. They found that this increase on 15-F_{2t}-IsoPs CSF levels was not correlated with the clinical signs of RRMS. Taken together, these results are representative of degenerative processes that occurs with or without the presence of inflammatory status. IsoPs seem to be highly produced in MS compared to other neurological diseases. Minghetti et al. (2000) reported 3-fold higher CSF levels of 15-F_{2t}-IsoPs in MS patients compared to a target group of patients with other neurological pathologies. The correlation of 15-F_{2t}-IsoPs levels was not statistically correlated; however, they were associated with a grade of neurologic-derived incapacity.

A research has reported CSF values of F_2-IsoPs in MS patients, their siblings and healthy unrelated controls [21]. The CSF content of F_2-IsoPs in siblings of MS patients was significantly higher than in healthy controls, but less than their brothers or sisters. Therefore, the increased oxidative stress in MS patient and their siblings could be a response to the environmental and genetic factors that might be partially responsible for MS pathogenesis.

Finally, studies in mice revealed the content of F_2-IsoPs and F_4-NeuroPs in experimental autoimmune encephalomyelitis [23, 24]. They suggested that, due to the low levels of DHA in white matter and myelin-forming oligodendrocytes, an increment on the F_2-IsoPs/F_4-NeuroPs product could be desirable for the development of MS disease.

4.2.2. Alzheimer's Disease

Alzheimer's disease (AD) is one of the major causes of dementia, defined by deposition of the peptide amyloid β (Aβ) and protein microtubule-associated in the SNC [25, 26]. The increase of cellular oxidative stress and the subsequent lipid peroxidation in AD is elevated, due to the interaction between Aβ and transition metals generating redox active ions [27]. The accumulation of oxidant compounds could be a cause of neurodegenerative procceses promoting the development of AD [28].

Many reports have shown the relationship between IsoPs and NeuroPs on the development of AD disease. Montine et al. (2002) reported increased values of F_2-IsoPs and F_4-NeuroPs in the postmortem ventricular fluid of AD patients. However, they found that the measurement of F_2-IsoPs from CSF could not be used as an early marker of dementia, due to the poor correlation of F_2-IsoPs with the different stages of the disease. Two IsoPs, 15-F_{2t}-IsoPs and 8,12-iso-iPF2alpha-VI, were found to have been increased in frontal and temporal lobes of the brain in AD patients, showing a positive correlation between them [30, 31].

NeuroPs have been also measured in AD patients. The research of Musiek et al. (2004) manifested the presence of F_4-NeuroPs in the human brain, especially in temporal lobes of brains from AD patients. The levels of F_4-NeuroPs measured was higher than the observed levels for F_2-IsoPs; however, this did not correlate with results obtained *in vitro* that showed a slight difference between these two oxidative stress markers [33]. Other researchers have shown the increase of F_4-NeuroPs in brain tissues. Nourooz-Zadeh et al. (1999) reported higher values of F_4-NeuroPs than F_2-IsoPs in temporal and occipital, but not parietal, lobes of brains in AD patients. In this line, Reich et al. (2001) stretched these results after the measurement of different families of IsoPs and NeuroPs (F-, D- and E-rings) in recent post-mortem brains from Ad patients. They found high content of total NeuroPs in brain regions involved with AD; however, IsoPs did suffer significant changes. Moreover, they found higher content of D/E-NeuroPs than F-NeuroPs in both, AD-affected and not AD-affected regions. This could mean that vulnerability in neurons of AD subjects is increased, and might be determined by specific factors of regions affected by AD.

The production of F_4-NeuroPs was also measured in AD patients and amnestic mild cognitive impairment subjects by Markesbery et al. (2005). They showed a significant increase of F_4-NeuroPs measured in parietal and occipital lobes in both cases, compared to control subjects.

The obtaining procces of CSF from the lumbar cistern is a safety method; however, it could be a stressful procedure for the patient. Moreover, the limitation of brain-focused studies has the obvious and inherent limitation of studying a patient post-mortem. Therefore, F_2-IsoPs have also been studied in other biological fluids such as urine and plasma. The results observed did not offer a uniform pattern about the production of F_2-IsoPs and their role on the development of AD. For example, Bohnstedt et al. (2003) did not show correlation between the urinary excretion of F_2-IsoPs and the development of AD disease. On the contrary, Kim et al. (2004) reported a statistical increase ofon the urinary excretion of F_2-IsoPs in patients with AD, concluding that F_2-IsoPs could be effective biomarkers of the disease. In plasma samples, Berr et al. (2011) described a poor correlation between the development of AD during 7 years and the increase of plasmatic F_2-IsoPs. Different studies using ELISA assays did not lead to detinitive statements. Feillet-Coudray et al. (1999) reported a similar urinary excretion of F_2-IsoPs in AD patients compared to a control group. On the contrary, Waddington et al. (1999) using ELISA kits, and Praticò et al. (2000) using a gas chromatography/mass spectrometric method, showed increased urinary and plasmatic levels of F_2-IsoPs in AD subjects compared to a control group.

Therefore, the relationship between the plasmatic concentration or the urinary excretion of F_2-IsoPs and the development of AD is still unclear.

4.2.3. Huntington's Disease

Huntington's disease (HD) is a neurodegenerative genetic disorder that affects muscle coordination and leads to mental decline and various behavioral symptoms. It is mainly characterized by an irregular increment of cytosine-adenineguanine repeat region at the 5'-end of a HTT gene on chromosome 4. The HTT gene provides instructions for making the huntingtin protein, so the mutation in the HTT gene causes HD [10]. The brain of HD patients has shown elevated free radical production and the subsequent oxidative stress injury, especially in the areas affected by HD. That increase of oxidative stress could be a cause or a consequence of the effects of HD [43, 44]. In HD patients, the overproduction of free radicals produced by mitochondrial dysfunction leads to in oxidative and nitrosative stress [45-47].

Moreover, the concentration of glutathione found in plasma samples from HD was significantly reduced, as reported by Stoy et al. (2005). Previous studies revealed the role of IsoPs in the development or cause of HD. Greco , Minghetti (2004), Greco et al. (2000) reported higher concentration of F_2-IsoPs in CSF from HD patients than from control subjects. Their results also showed no differences between men and women, or a correlation with F_2-IsoPs and the duration of the disease. Finally, they also deduced that the oxidative damage occurred during HD is not similar in all the phases of the disease, due to overlap between groups. Other groups have done researchs in transgenic HD mice. They suggested that increased immunostaining for malondialdehyde, 4-hydroxynonenal, and 15-F_{2t}-IsoPs [46, 49].

Therefore, the investigation about HD and IsoPs production is at its earliet stages and thusmore studies are needed.

4.2.4. Creutzfeldt-Jakob Disease

Creutzfeldt-Jakob disease (CJD) is a degenerative brain disorder that leads to dementia and, ultimately, death. CJD is one of the most known human transmissible spongiformencephalopathies, caused by prion infection. Symptoms of CJD are similar to those of other dementia-like brain disorders, such as AD. However, CJD progression is much more rapid than other similar pathologies [10]. Neuropathological basis of CJD is similar to other diseases such as AD (Aβ) or HD (huntingtin), due to the amyloid deposition of the pathological form of a cellular protein. The CJD prion endorses refolding of native proteins into the diseased state. The number of misfolded protein molecules will increase exponentially, leading to a large quantity of insoluble protein in affected cells. This mass of misfolded proteins disrupts neuronal cell function and causes cell death. Mutations within the gene for the prion protein can cause a misfolding of the dominantly alpha helical regions into beta-pleated sheets. This change in conformation disables the ability of the protein to undergo digestion. Once the prion is transmitted, the defective proteins invade the brain and are produced in a self-sustaining feedback loop. The accumulation of the pathological prion protein leads to several pathogenetic mechanisms, concluding in spongiform degeneration [33].

The biological activity of cellular prion protein is still unknown. The main described function is its cooper binding capacity. This fact leads to antioxidant properties, owing to its superoxide dismutase- (SOD-) like activity, and the maintaining of cellular homeostasis. That fact was studied by Kralovicova et al. (2009) and Wong et al. (2001) who reported that cells expressing higher levels of prion protein were more resistant to oxidative stress and higher levels of peroxidation products in brain tissues from knockout mice, respectively. In turn, Arlt et al. (2002) observed minor concentration of PUFAs in CSF from CJD patients compared with a control group, supporting the theory of the susceptibility of CJD patients to oxidative stress and correlating those levels of PUFAs with pathogenesis of this CJD.

The formation of IsoPs has also been describedwithin the current scientific literature. Minghetti et al. (2002) observed an increase of F_2-IsoPs levels in CSF from CJD patients. CSF levels of 15-F_{2t}-IsoP were about 2.5-fold higher in CJD patients; however, they reported that lipid peroxidation and prostaglandin synthesis (measured by the wuantification of PGE_2) were not correlated. Therefore, the inflammation might be more relevant than oxidative stress to the pathogenesis of CJD [33, 53].

4.2.5. Epilepsy

Despite the advances done in the field of epilepsy, the cause of most cases of epilepsy occurrence is still unknown. However, an increment on the oxidant status of the organism has been related to the development of epilepsy [54]. That fact has been noted by the scientific community, finding an association between the increment of oxidant species and the development of neurological diseases [55, 56].

The only approach to the formation of NeuroPs and dihomo-IsoPs in epileptic patients was done by Medina et al. (2015). They found an increase in the urinary excretion of different F_4-NeuroPs and F_2-dihomo-IsoPs in epilectic, demonstrating higher oxidative stress derived from this condition. Finally, they proposed 17-F_{2t}-Dihomo-IsoP, *ent*-7(R)-7-F_{2t}-Dihomo-IsoP and *ent*-7-*epi*-7-F_{2t}-Dihomo-IsoP as a propising biomarkers of epilepsy.

4.2.6. Brain Injury

The high oxygen requirements of the brain and its PUFAs composition is a point for the development of neurodegenerative disorders; however, it also makes the brain vulnerable to oxidative acute insult [58]. Two acute brain-related diseases such as aneurysmal subarachnoid hemorrhage (aSAH) and traumatic brain injury (TBI) are related to overproduction of oxidant species. In a clinical research, Corcoran et al. (2011) reported increased levels of IsoFs and F_4-NeuroPs in CSF from aSAH and TBI patients compared with control groups. Moreover, the analysis of CSF from aSAH patients, but not TBI patients, also reported increased values of F_2-IsoPs. Similar to neurodegenerative disorders, F_4-NeuroPs concentration measured in CSF could be a more specific biomarker of acute brain injury than F_2-IsoPs (due to DHA is the major polyunsaturated fatty acid in the brain). That fact was supported by the reports of Hsieh et al. (2009), who established a correlation between the increase of F_4-NeuroPs and poor neurological outcomes, concluding that F_4-NeuroPs measured in CSF are more specific biomarkers of oxidative stress in SNC than F_2-IsoPs.

4.2.7. Pre-Eclampsia

Pre-eclampsia is a deadly disorder that occurs during gestation that is dangerous for both the mother and the baby, which pathogeny could be affected by oxidative stress. It has previously been reported that pre-eclampsia could increase the plasmatic concentration of F_2-IsoPs [61]. The same authors reported an increment of plasmatic and cord blood F_4-NeuroPs and IsoFs, but not F_2-IsoPs, in pregnant women affected by pre-eclampsia [62]. Both cord blood concentration of F_4-NeuroPs and IsoFs were higher in babies from Pre-eclampsia affected mothers; however, IsoFs concentration from cord blood was 5-fold higher than plasmatic concentration found in plasma samples from their mothers. Authors theorized that this increase compared to maternal plasma samples makes relevant the importance of transition from a relatively low intrauterine oxygen background to a significantly higher extrauterine oxygen environment.

4.2.8. Amyotrophic Lateral Sclerosis

Apart from being a multifactorial and complex disease, amyotrophic lateral sclerosis (ALS) can be defined by progressive degeneration of the motor neurons. Similar to other neurodegerative diseases, the main chareacteristic of the ALS is the accumulation of a misfolded protein (copper/zinc SOD) in neural tissue [63]. The deposition of the copper/zinc SOD leads to the development of ALS, characterized by the increase of proinflammatory molecules, astrogliosis, and microglial activation in spinal cord tissue [10].

The pathogeny of ALS is quite related to oxidative stress and motor neuron deterioration, mainly due to an increase in the inflammation procceses which leads to neuron death [64]. Numerous clinical studies shown the relationship between ALS and different biomarkers of oxidative stress..The urinary excretion of 15-F_{2t}-IsoP increased in ALS patients compared to control patients, unrelated to the age of participants. The urinary excretion of 8-oxodG seemed to follow the same trend as the observed for 15-F_{2t}-IsoP [65]. However, the formation of carbonyl groups from proteins did not increase in ALS patients. Therefore, both 8-oxodG and 15-F_{2t}-IsoP but not carbonyl groups seem to be useful as biomarkers of ALS [65, 66].

Scientific literature proposes the inhibition of COX-2 as benefitial, due to its capacity to alter diverse pathways leading to neuronal death in ALS patients. In fact, COX-2 could exert a determinant role in the inflammation and in glutamate-dependent neurotoxicity, which determines the possible association of COX-2 in ALS pathogenesis [67]. In ALS patients, it was observed the increase of PGE_2 and COX-2. The COX-2 expression is localized to both neurons and glial cells in the postmortem spinal cord of ALS patients, being more associated with astrocytes more than in glial cells. Moreover, diverse regions like the spinal cord dorsal, ventral horns or dorsal root ganglia have been shown to contain significant quantities of COX-2 in neurons [67]. Finally, 15d-PGJ_2 (a prostanoid derived dehydration and isomerization of PGD_2) triggers the nuclear peroxisome proliferator activated receptor γ, a critical transcription factor involved in adipocyte and monocyte differentiation [68].

4.3. Cancer

Increased oxidative stress may be involved in carcinogenesis through the induction of gene mutations that result from cell injury and the effects on signal transduction and transcription factors. The kind of injury depends on both the type and quantity of reactive oxygen species. Moreover, the factors involved could be the production of free radicals by tumor cells, activation of neutrophils and macrophages or the disfunction of endogenous antioxidants controlling the redox status. Therefore, lipid peroxidation products have been studied and are related to the development of different cancers or tumor growing.

The increment of the urinary excretion of F_2-IsoPs was related with increased risk of hepatocellular carcinoma by [69]. In turn, Rossner et al. (2006) reported increased levels of F_2-IsoPs in patients with breast cancer. The differences in the urinary excretion of F_2-IsoPs between gender was accomplished by Epplein et al. (2009), who reported a correlation between urinary F_2-IsoPs and risk of lung cancer in men but not women. Sequentially, prostate and gastric cancer has been related to the increase of urinary F_2-IsoPs [72-74]. Contrary to what was commented on before, Siamakpour-Reihani et al. (2012) did not found a strong correlation between the occurrence of adenoma and the increment of urinary excretion of four metabolites of F_2-IsoPs.

These results highlight the relevance of the oxidative injury in relation to cancer. However, the limitation in the analysis of long-term diseases is the lack of a consistent monitoring to allow us from knowing the evolution of the disease in different stages and to clarify if oxidative stress leads or results from cancer.

It that sense, Dai et al. (2009) reported the results of the largescale nested Shanghai Women's Health Study (SWHS). That study was comprised of almost 75,000 Chinese women between the ages of 40 and 70 years old. They measured urinary F_2-IsoP and the metabolite 2,3-dinor- 5,6-dihydro-15-F_{2t}-IsoP as biomarkers of breast cancer risk. Their results showed that the urinary excretion of 2,3-dinor- 5,6-dihydro-15-F_{2t}-IsoP and total F_2-IsoP was related to a reduced risk of breast cancer among overweight and obese women. This relationship was higher in women with higher BMI and more marked for 2,3-dinor- 5,6-dihydro-15-F_{2t}-IsoP than for total F_2-IsoP. Finally, other studies have also notified the relationship of cancer with both BMI and oxidative stress. Thompson , Ambrosone (2000) found that overweight and obese women with high urinary excretion of F_2-IsoPs, were at an increased risk of breast cancer.

These conclusions point out the role of the BMI in the development of breast cancer mediated by free oxidative stress. Therefore, the role of fat tissues in the development of cancer and oxidative stress-related seems to be important.

4.4. Obesity

The occurance of obesity has increased dramatically over the last several years, becoming a recognized healt problem around the world. Obesity, characterized by an excess of body fat, is highly prevalent in the United States as well as globally. Obesity is characterized by an increment of corporal fat and the subsequent general inflammation. Moreover, it is a complex condition with many causal contributors and can result in many complications for the individuals affected by this condition. Both morbidity and mortality increase with excessive body weight [78]. There are numerous studies reporting increments of plasmatic and urinary levels of F_2-IsoPs in obese adults [79-85]. The urinary excretion and plasmatic concentration of F_2-IsoPs have been previously correlated with body mass index (BMI), waist circumference (WC), visceral fat area, and percent body fat [76, 86-101]. More recently, differences in formation of F_2-IsoPs has been reported between races. Il'yasova et al. (2012) found high and BMI-correlated urinary levels of F_2-IsoPs and 2,3-dinor-15-F_{2t}-IsoP in non-Hispanic whites and Hispanic whites patients but not in African Americans. However, a clinical study carried out with kids between

8-17 years old, reported a correlation between plasmatic concentrations of F_2-IsoPs and percent body fat as well as truncal fat. However, there were no differences regarding BMI and race [102]. Moreover, a study of Chinese women reported urinary 2,3-dinor-5,6-dihydro-15-F_{2t}-IsoP but not urinary F_2-IsoPs correlation with increased BMI [103].

Those previous results indicate that obese populations are further affected by oxidative stress than healthy populations, judging by the measurement of IsoPs. Moreover, it is important to analyze various biological fluids or tissues, in order to ensure a complete measurement of the redox status.

4.5. Diet

The formation of F_2-IsoPs regarding the dietary intake of antioxidants has lead to a wide research field over the last few years. However, the results obtained by the different studies are not consistent. This fact is produced because of the differences of the study design, such as different doses of antioxidants, length of the supplementation, different biological fluids analyzed, and the metabolite of IsoPs chosen for the measurement of oxidative stress [3, 104].

Specifically, vitamin E-related clinical researches would benefit from further investigation reporting standardized values of vitamin E for the correct maintenance of health. Roberts et al. (2007) established that low values of vitamin E achieve a significant reduction of lipid peroxidation, reporting that daily intake of 1600 IU of vitamin E during 8 weeks was able to decrease the plasmatic concentration of F_2-IsoPs in healthy humans. Other studies have shown the changes on F_2-IsoPs formation after dietetic intervention in different pathologies. Block et al. (2008) reported that 1000 mg/day intake of vitamin C and 800 IU/day intake of vitamin E were able to reduce plasmatic concentration of F_2-IsoPs in obese subjects (22% and 10% respectively). However, neither the intake of vitamin C or vitamin E, were able to reduce plasma F_2-IsoPs in low baseline than those found in obese population. Moreover, the supplementation with vitamin E acetate leads to the decrease of F_2-IsoPs measured in brochoalveolar lavage fluid of atopic asthmatics and control subjects [107]. In a study of Chinese women using vitamin E supplement, Dorjgochoo et al. (2012) showed a diminution in the urinary excretion of 2,3-dinor-5,6-dihydro-15-F_{2t}-IsoP, but not F_2-IsoPs. They found an inverse correlation between the plasmatic contents of antioxidants, as β-carotenes or lycopenes, and 2,3-dinor-5,6-dihydro-15-F_{2t}-IsoP. On the contrary, they found a positive correlation between plasmatic concentration of β-, γ-, and δ- tocopherols and urinary excretion of 2,3-dinor-5,6-dihydro-15-F_{2t}-IsoP.

Other dietary antioxidants have also been shown to influence the formation of IsoPs formation. For example vitamin E, and the oral intake of other antioxidants can especially reduce the high basal levels of isoprostanes, as is seen in obesity or asthma, more than in normal healthy populations. In clinical models, Espey et al. (2011) showed that vitamin C was able to decrease pancreatic cancer progression. That vitamin was also studied intravenous in a clinical study, being able to decrease the plasmatic levels of F_2-IsoPs after the infusion.

The intake of berries (blackberries, strawberries, blueberries, blackcurrants, and cherries) has especially demonstrated a reduction in the formation of IsoPs. The intake of tart cherry juice reduced the formation of F_2-IsoPs, after the increase observed in blood pressure cuff inflation patients [109]. A clinical study with athletes reported that the daily intake of 250 g of blueberries during 6 weeks was able to reduce the plasmatic increased production of F_2-IsoPs, produced after 150 minutes of training [110]. The effect of the comsumption of blackcurrant was also studied. Khan et al. (2014) demonstrated the healthy benefits of the intake of a blackcurrant juice drink by the reduction of basal plasmatic concentration of F_2-IsoPs in healthy population.

The entirety of diet behaviors, controled by dietary intake surveys, has also been under study. A long-time dietary measurement survey during 20 years with more than 5000 young and middle-aged volunteers, reported than the intake of fruits and vegetables apart from a low consumption of red meat were associated with lower levels of plasma F_2-IsoPs [112]. On the contrary, the intake of high quantity of red meat was found to be associated with an increment of plasmatic F_2-IsoPs. The intake of fruits and vegetables reported conflicting results in other studies. In a survey with 258 premenopausal subjects, the volunteers following the requirement of at least five servings of fruits and vegetables (marked by the 5 A Day for Better Health Program) showed minor plasmatic levels of F_2-IsoPs [113]. However, in another study comparing pre and postmenopausal women, that diminution in F_2-IsoPs levels was not found in urine [114]. That fact makes even more relevant the measurement of F_2-IsoPs in different biological fluids, in order to counteract the differences observed between matrices.

Finally, the intake of other bioactive compounds such as omega-3 PUFAs (DHA and EPA) also showed antioxidant properties, reducing the formation of F_2-IsoPs [85, 115-117].

4.6. Genetic Disorders

Repeated oxidative injury caused by free radicals can be developed in different genetic disorders, characterized by the mutation of a single gene or an entire chromosome. The formation of F_2-IsoPs has been deeply studied in Rett syndrome (RTT) [15, 117-119]. However, F_2-IsoPs were measured in other diseases such as autism-spectrum disorders [120, 121], Smith-Lemli-Opitz Syndrome (SLOS) [122], sickle cell anemia [123], cystic fibrosis [124], in addition to various inborn errors of metabolism [125]. The results reported by Perrone et al. (2007) showed an increase of F_2-IsoPs in the amniotic fluid of expectant mothers of children with Down syndrome.

As mentioned above, oxidative stress and IsoPs are extensively studied in RTT patients. RTT is a neurological disease characterized by a genetic defect in young females involving the methyl-CpG binding protein 2 (MeCP2) [15]. RTT is characterized by deficitary cognitive capacity besides some similarities with autistic patients. The different scientific reports have established higher plasmatic values of F_2-IsoPs, F_4-NeuroPs and F_2-dihomo-IsoPs in RTT patients [15, 118, 119, 127]. As a whole, F_4-NeuroPs are related with the severity of RTT; meanwhile, the values of F_2-dihomo-IsoPs seem to increase in the early stages of the disease. Signorini et al. (2014) developed a clinical survey about the intake of oral supplementation with 20–40 mg/kg marine fish oil. They reported a reduction in plasmatic F_4-NeuroPs of the 35% of the patients after 12 months of treatment.

4.7. Biological Role of Isoprostanes

IsoPs, NeuroPs or dihomo-IsoPs are biomarkers of lipid peroxidation, but can exert other biological functions. IsoPs (as reviewed by Galano et al. (2013)), can especially exert vasoconstrictor activity [129], moduling the activity of platelets [130] or monocyte adhesion [131], and induce proliferation of endothelial and smooth muscle cells [132, 133].

The quite similar structure of IsoPs regarding PGs has focused the investigation of their bioactivity on the cardiovascular system. Different activity of IsoPs as free molecules has been described by the scientific literature, and the most studied IsoPs is 15-F_{2t}-IsoP.

A couple of years ago, Bauer et al. (2014) revised the effects of IsoPs on the cardiovascular system. Concisely, 15-F_{2t}-IsoP can act as a potent vasoconstrictor, being able to control platelet activity, inhibit angiogenesis, and promote atherosclerosis by stimulating the adhesion of monocytes and endothelial neutrophils. These biological activities of 15-F_{2t}-IsoP are determined by the interaction with the thromboxane receptor (TP).

Similarly, 15-E_{2t}-IsoPs also mediates its functions through the TP receptor. Both 15-E_{2t}-IsoPs and 15-F_{2t}-IsoPs have exerted biological activity through other PG receptors (PGE_2 receptor (EP) and 15-F_2-IsoP (FP)) [134-144].

Notably, the activity of the molecules may vary depending on the receptor to which it binds. For example, while 15-E_{2t}-IsoPs and 15-F_{2t}-IsoPs are vasoconstrictors through TP receptors, they can both induce vasodilation through the EP receptor. These opposing effects have been particularly studied in the lungs [142]. In addition, Chen et al. (2012) have reported both, vasoconstrictor and vasodilator, 15-E_{2t}-IsoPs and 15-F_{2t}-IsoPs in the arterial duct Botalli, a central vascular shunt shortly constricted after birth [142]. These authors have shown that the ductus arteriosus of Botalli contracts in response to both 15-E_{2t}-IsoPs and 15-F_{2t}-IsoPs, in a concentration-dependent way. Moreover, this effect could be reversed by administering an antagonist of the TP receptor. Interestingly, when TP receptors clatter before the administration of IsoPs, both 15-E_{2t}-IsoPs and 15-F_{2t}-IsoPs lead to increased vasodilation. This fact is presumably produced due to the activation of the EP_4 receptor. They can act as a potent vasoconstrictor dose-dependent in chicken embryos, acting through the TP receptor and modified by the presence of H_2O_2 [145, 146]. More studies about 15-F_{2t}-IsoP and 15-E_{2t}-IsoP revealed the importance of the production of hyperalgesia, an increased sensitivity to pain in rats [147]. Finally, other related work has found that IsoPs levels are higher in newborns than in older infants [148].

The latest research on the biological effects of isoprostanes allow us to understand the effect of an oxidative attack as it relates to the increased inflammation at a systemic level. Therefore, it can be a very useful tool to neutralize the potentially harmful consequences of increased oxidative stress in the organism.

Many pontential uses regarding individual compounds can be determined [128]. Some of them are discussed then.

4.7.1. Isops and Cardiovascular System

The effects of 15-F_{2t}-IsoP have been previously approached above. The most important action of 15-F_{2t}-IsoP is the vasomotor action, investigated in isolated human saphneous and umbilical veins, mammary arteries, bronchia arteries, pulmonar arteries and placental and maternal vessels [149]. The vasoconstriction observed in 15-F_{2t}-IsoP is not standarized to other IsoPs series. In fact, 5-F_2-IsoP are not able to produce the constriction observed in 15-IsoPs series [150].

Other biological functions have been observed for IsoPs. IsoPs not only are able to produce vasoconstriction and platelet activation, but also enhance vascular reperfusion injury next to myocardial infarction [151]. Moreover, 15-F_{2t}-IsoP stimulate endothelial cell migration, even in short concentration. However, other IsoPs series from E-IsoPs, A-IsoPs (even some metabolites from F-IsoPs) series inhibit the migration and tube formation of endothelial cells vía VEGF. mentioned in Galano et al. (2013), these efects are "abolished either by TP blockade or alternatively by short hairpin RNA-mediated knock down of the TP".

As a conclussion, the role of 15-F_{2t}-IsoP and other IsoPs series can be determinant on the homeostasis of cardiovascular system, due to the TP adjustement.

4.7.2. IsoPs in Retina

The tissues that make up the retina are constanly expossed to external light and suffer from consquent oxidation due to the chemical processes occuring in the retina. Moreover, the retina is formed by high quentities of polyunsaturated fatty acids, which made it quite susceptible to oxidative injury [152].In fact, oxidative stress is the main cause of many diseases like glaucoma, leading to the formation of free radicals in the retina tissue [153, 154].

The functions of 15-F_{2t}-IsoP in retina is mainly comprised of the neurotransmission of diverse stimuli [153]. Nevertheles, the functions of other series as 5-F_2-IsoP have not been described extensively. However, it is believed that this could affect the release of glutamate, as previously studied in bovine retina tissues in vitro [155]. The capacity of glutamate to be a major neurotransmitter excitator, apart from its involvement in neuronal excitotoxicity, makes extremely relevant the role of 5-F_2-IsoP to decrease the release of neurotransmitters that have noteworthy pathophysiological consequences in the retina tissues. Therefore, the determination of 5-F_2-IsoP and 15-F_{2t}-IsoP could be a robust way to determine the possible degeneration of retina tissues, making ocular neuropathies promising biomarkers.

4.7.3. IsoPs Implication on Anti Arrhythmic Processes

It is established that diets with high intake of Ω-3 polyunsaturated fatty acids exert cardioprotective effects, mainly due to EPA and DHA [156]. Mozaffarian , Wu (2011) demonstrated that both DHA and EPA are able to exert a marked effect by reducing the appearance of sudden cardiac death in the following months of a cardiac infarction. That fact was atributted to the reduction on the onset of arrhythmias and systolic cardiac failure. Moreover, Ω-3 polyunsaturated fatty acids can also exert antiarrhythmiceffects as determined in animal models [158].

The oxydized products of EPA and DHA are supposed to be able to have the same effects as their primary polyunsaturated fatty acids [158]. For example, F_4-NeuroPs can partially exert the effects described for DHA [159]. Finally, Galano et al. (2013) confirmed the relationship of many F_2-IsoPs, F_3-IsoPs and F_4-NeuroPs on isoprenaline-induced arrhythmias of isolated ventricular mice cardiac cells. Various F_4-NeuroPs especially demonstrated antiarrhythmic capacity (IC50 $\approx$ 100 nM).

REFERENCES

[1] Popescu, Bogdan O. 2013. "Triggers and effectors of oxidative stress at blood-brain barrier level: relevance for brain ageing and neurodegeneration." *Oxidative medicine and cellular longevity* no. 2013.

[2] Santos, Renata, Carmen Ruiz de Almodóvar, Anne-Laure Bulteau, Cláudio M Gomes. 2013. "Neurodegeneration, neurogenesis, and oxidative stress." *Oxidative medicine and cellular longevity* no. 2013.

[3] Halliwell, B., C. Y. J. Lee. 2010. "Using isoprostanes as biomarkers of oxidative stress: Some rarely considered issues." *Antioxidants and Redox Signaling* no. 13 (2):145-156. doi: 10.1089/ars.2009.2934.

[4] Barrera, Giuseppina. 2012. "Oxidative stress and lipid peroxidation products in cancer progression and therapy." *ISRN oncology* no. 2012.

[5] Oger, C., V. Bultel-Poncé, A. Guy, T. Durand, J. M. Galano. 2012. "Total synthesis of isoprostanes derived from adrenic acid and EPA." *European Journal of Organic Chemistry* (13):2621-2634. doi: 10.1002/ejoc.201200070.

[6] Milne, Ginger L, Qi Dai, L Jackson Roberts. 2015. "The isoprostanes—25 years later." *Biochimica et Biophysica Acta (BBA)-Molecular and Cell Biology of Lipids* no. 1851 (4):433-445.

[7] Kadiiska, M. B., B. C. Gladen, D. D. Baird, D. Germolec, L. B. Graham, C. E. Parker, A. Nyska, J. T. Wachsman, B. N. Ames, S. Basu, N. Brot, G. A. FitzGerald, R. A. Floyd, M. George, J. W. Heinecke, G. E. Hatch, K. Hensley, J. A. Lawson, L. J. Marnett, J. D. Morrow, D. M. Murray, J. Plastaras, L. J. Roberts Ii, J. Rokach, M. K. Shigenaga, R. S. Sohal, J. Sun, R. R. Tice, D. H. Van Thiel, D. Wellner, P. B. Walter, K. B. Tomer, R. P. Mason, J. C. Barrett. 2005. "Biomarkers of Oxidative Stress Study II: Are oxidation products of lipids, proteins, and DNA markers of CCl4 poisoning?" *Free Radical Biology and Medicine* no. 38 (6):698-710. doi: http://dx.doi.org/10.1016/j.freeradbiomed.2004.09.017.

[8] Il'Yasova, D., I. Spasojevic, F. Wang, A. A. Tolun, K. Base, S. P. Young, P. K. Marcom, J. Marks, G. Mixon, R. DiGiulio, D. S. Millington. 2010. "Urinary biomarkers of oxidative status in a clinical model of oxidative assault." *Cancer Epidemiology Biomarkers and Prevention* no. 19 (6):1506-1510. doi: 10.1158/1055-9965.EPI-10-0211.

[9] Barden, A. E., E. Mas, K. D. Croft, M. Phillips, T. A. Mori. 2014. "Minimizing artifactual elevation of lipid peroxidation products (F 2-isoprostanes) in plasma during collection and storage." *Analytical biochemistry* no. 449 (1):129-131. doi: 10.1016/j.ab.2013.12.030.

[10] Miller, E., A. Morel, L. Saso, J. Saluk. 2014. "Isoprostanes and neuroprostanes as biomarkers of oxidative stress in neurodegenerative diseases." *Oxidative Medicine and Cellular Longevity* no. 2014.

[11] Cappellano, G., M. Carecchio, T. Fleetwood, L. Magistrelli, R. Cantello, U. Dianzani, C. Comi. 2013. "Immunity and inflammation in neurodegenerative diseases." *Am J Neurodegener Dis* no. 2 (2):89-107.

[12] Almer, G., P. Teismann, Z. Stevic, J. Halaschek-Wiener, L. Deecke, V. Kostic, S. Przedborski. 2002. "Increased levels of the pro-inflammatory prostaglandin PGE2 in CSF from ALS patients." *Neurology* no. 58 (8):1277-1279.

[13] Farooqui, T., A. A. Farooqui. 2011. "Lipid-mediated oxidative stress and inflammation in the pathogenesis of Parkinson's disease." *Parkinson's Disease*. doi: 10.4061/2011/247467.

[14] Guest, J., R. Grant, T. A. Mori, K. D. Croft. 2014. "Changes in oxidative damage, inflammation and [NAD(H)] with age in cerebrospinal fluid." *PLoS ONE* no. 9 (1).

[15] Durand, T., C. De Felice, C. Signorini, C. Oger, V. Bultel-Poncé, A. Guy, J. M. Galano, S. Leoncini, L. Ciccoli, A. Pecorelli, G. Valacchi, J. Hayek. 2013. "F2-Dihomo-isoprostanes and brain white matter damage in stage 1 Rett syndrome." *Biochimie* no. 95 (1):86-90.

[16] Minghetti, L., R. Salvi, M. Lavinia Salvatori, M. Antonietta Ajmone-Cat, C. De Nuccio, S. Visentin, V. Bultel-Poncé, C. Oger, A. Guy, J. M. Galano, A. Greco, A. Bernardo, T. Durand. 2014. "Nonenzymatic oxygenated metabolites of α-linolenic acid B 1- and L1-phytoprostanes protect immature neurons from oxidant injury and promote differentiation of oligodendrocyte progenitors through PPAR-γ activation." *Free Radical Biology and Medicine* no. 73:41-50.

[17] Wong, B. S., D. R. Brown, T. Pan, M. Whiteman, T. Liu, X. Bu, R. Li, P. Gambetti, J. Olesik, R. Rubenstein, M. S. Sy. 2001. "Oxidative impairment in scrapie-infected mice is associated with brain metals perturbations and altered antioxidant activities." *Journal of Neurochemistry* no. 79 (3):689-698.

[18] Peterson, Lisa K, Robert S Fujinami. 2007. "Inflammation, demyelination, neurodegeneration and neuroprotection in the pathogenesis of multiple sclerosis." *Journal of neuroimmunology* no. 184 (1):37-44.

[19] Miller, Elżbieta, Małgorzata Mrowicka, Katarzyna Malinowska, Jerzy Mrowicki, Joanna Saluk-Juszczak, Józef Kędziora. 2011. "Effects of whole-body cryotherapy on a total antioxidative status and activities of antioxidative enzymes in blood of depressive multiple sclerosis patients." *The World Journal of Biological Psychiatry* no. 12 (3):223-227.

[20] Miller, Elżbieta, Anna Walczak, Joanna Saluk, Michał B Ponczek, Ireneusz Majsterek. 2012. "Oxidative modification of patient's plasma

proteins and its role in pathogenesis of multiple sclerosis." *Clinical Biochemistry* no. 45 (1):26-30.

[21] Mattsson, Niklas, Sara Haghighi, Oluf Andersen, Yuemang Yao, Lars Rosengren, Kaj Blennow, Domenico Praticò, Henrik Zetterberg. 2007. "Elevated cerebrospinal fluid F2-isoprostane levels indicating oxidative stress in healthy siblings of multiple sclerosis patients." *Neuroscience letters* no. 414 (3):233-236.

[22] Minghetti, Luisa, Anita Greco, Franco Cardone, Maria Puopolo, Anna Ladogana, Susanna Almonti, Colm Cunningham, V Hugh Perry, Maurizio Pocchiari, Giulio Levi. 2000. "Increased brain synthesis of prostaglandin E2 and F2-isoprostane in human and experimental transmissible spongiform encephalopathies." *Journal of Neuropathology & Experimental Neurology* no. 59 (10):866-871.

[23] Greco, Anita, Luisa Minghetti. 2004. "Isoprostanes as biomarkers and mediators of oxidative injury in infant and adult central nervous system diseases." *Current neurovascular research* no. 1 (4):341-354.

[24] Ljubisavljevic, Srdjan, Ivana Stojanovic, Dusica Pavlovic, Maja Milojkovic, Dusan Sokolovic, Ivana Stevanovic, Aleksandar Petrovic. 2013. "Suppression of the lipid peroxidation process in the CNS reduces neurological expression of experimentally induced autoimmune encephalomyelitis." *Folia Neuropathologica* no. 51 (1):51-57.

[25] Mondragón-Rodríguez, Siddhartha, George Perry, Xiongwei Zhu, Jannic Boehm. 2012. "Amyloid beta and tau proteins as therapeutic targets for Alzheimer's disease treatment: rethinking the current strategy." *International Journal of Alzheimer's Disease* no. 2012.

[26] Savelieff, Masha G, Sanghyun Lee, Yuzhong Liu, Mi Hee Lim. 2013. "Untangling amyloid-β, tau, and metals in Alzheimer's disease." *ACS chemical biology* no. 8 (5):856-865.

[27] Ellis, Grace, Erica Fang, Mukesh Maheshwari, Emily Roltsch, Leigh Holcomb, Danna Zimmer, Daniel Martinez, Ian VJ Murray. 2010. "Lipid oxidation and modification of amyloid-β (Aβ) in vitro and in vivo." *Journal of Alzheimer's Disease* no. 22 (2):593.

[28] Pritchard, Susanne M, Philip J Dolan, Alisa Vitkus, Gail VW Johnson. 2011. "The toxicity of tau in Alzheimer disease: turnover, targets and potential therapeutics." *Journal of cellular and molecular medicine* no. 15 (8):1621-1635.

[29] Montine, Thomas J, Joseph F Quinn, Dejan Milatovic, Lisa C Silbert, Theresa Dang, Stephanie Sanchez, Erin Terry, L Jackson Roberts, Jeffrey A Kaye, Jason D Morrow. 2002. "Peripheral F2-isoprostanes

and F4-neuroprostanes are not increased in Alzheimer's disease." *Annals of Neurology* no. 52 (2):175-179.

[30] Yao, Y, V Zhukareva, S Sung, CM Clark, J Rokach, VM-Y Lee, JQ Trojanowski, D Pratico. 2003. "Enhanced brain levels of 8, 12-iso-iPF2α-VI differentiate AD from frontotemporal dementia." *Neurology* no. 61 (4):475-478.

[31] Praticò, Domenico, Virginia M-Y Lee, John Q Trojanowski, Joshua Rokach, Garret A Fitzgerald. 1998. "Increased F2-isoprostanes in Alzheimer's disease: evidence for enhanced lipid peroxidation in vivo." *The FASEB Journal* no. 12 (15):1777-1783.

[32] Musiek, Erik S, Jin K Cha, Huiyong Yin, William E Zackert, Erin S Terry, Ned A Porter, Thomas J Montine, Jason D Morrow. 2004. "Quantification of F-ring isoprostane-like compounds (F 4-neuroprostanes) derived from docosahexaenoic acid in vivo in humans by a stable isotope dilution mass spectrometric assay." *Journal of Chromatography B* no. 799 (1):95-102.

[33] Greco, Anita, Luisa Minghetti, Giulio Levi. 2000. "Isoprostanes, novel markers of oxidative injury, help understanding the pathogenesis of neurodegenerative diseases." *Neurochemical research* no. 25 (9-10):1357-1364.

[34] Nourooz-Zadeh, Jaffar, Edwin HC Liu, B Yhlen, Erik E Änggåard, Barry Halliwell. 1999. "F4-Isoprostanes as Specific Marker of Docosahexaenoic Acid Peroxidation in Alzheimer's Disease." *Journal of neurochemistry* no. 72 (2):734-740.

[35] Reich, Erin E, William R Markesbery, L Jackson Roberts, Larry L Swift, Jason D Morrow, Thomas J Montine. 2001. "Brain regional quantification of F-ring and D-/E-ring isoprostanes and neuroprostanes in Alzheimer's disease." *The American journal of pathology* no. 158 (1):293-297.

[36] Markesbery, William R, Richard J Kryscio, Mark A Lovell, Jason D Morrow. 2005. "Lipid peroxidation is an early event in the brain in amnestic mild cognitive impairment." *Annals of Neurology* no. 58 (5):730-735.

[37] Bohnstedt, Kristina Claeson, Bo Karlberg, Lars-Olof Wahlund, Maria Eriksdotter Jönhagen, Hans Basun, Staffan Schmidt. 2003. "Determination of isoprostanes in urine samples from Alzheimer patients using porous graphitic carbon liquid chromatography–tandem mass spectrometry." *Journal of Chromatography B* no. 796 (1):11-19. doi: http://dx.doi.org/10.1016/S1570-0232(03)00600-7.

[38] Kim, Kyung Mee, Byung Hwa Jung, Ki-Jung Paeng, Inseong Kim, Bong Chul Chung. 2004. "Increased urinary F2-isoprostanes levels in the patients with Alzheimer's disease." *Brain Research Bulletin* no. 64 (1):47-51. doi: http://dx.doi.org/10.1016/j.brainresbull.2004.04.016.

[39] Berr, Claudine, Tasnime Akbaraly, Françoise Michel, Anne-marie Dupuy, Laure Anne Gutierrez, Florence Portet, Melanie Le Goff, Luc Letenneur, Jean-Paul Cristol, Pascale Barberger-Gateau. 2011. "Plasma F2 isoprostanes and dementia : Results from a longitudinal study." *Alzheimer's & Dementia* no. 7 (4, Supplement):S348. doi: http://dx.doi.org/10.1016/j.jalz.2011.05.1009.

[40] Feillet-Coudray, Christine, Rémi Tourtauchaux, Michel Niculescu, Edmond Rock, Igor Tauveron, Marie-Cecile Alexandre-Gouabau, Yves Rayssiguier, Isabelle Jalenques, Andrzej Mazur. 1999. "Plasma levels of 8-epiPGF2α, an in vivo marker of oxidative stress, are not affected by aging or Alzheimer's disease." *Free Radical Biology and Medicine* no. 27 (3):463-469.

[41] Waddington, E, K Croft, R Clarnette, T Mori, R Martins. 1999. "Plasma F-2-isoprostane levels are increased in Alzheimer's disease: evidence of increased oxidative stress in vivo." *Alzheimers reports* no. 2 (5):277-282.

[42] Praticò, Domenico, Christopher M Clark, Virginia M-Y Lee, John Q Trojanowski, Joshua Rokach, Garret A FitzGerald. 2000. "Increased 8, 12-iso-iPF2α-VI in Alzheimer's disease: Correlation of a noninvasive index of lipid peroxidation with disease severity." *Annals of Neurology* no. 48 (5):809-812.

[43] Sánchez-López, Fernando, Inmaculada Tasset, Eduardo Agüera, Montserrat Feijóo, Ricardo Fernández-Bolaños, Francisco M Sánchez, María C Ruiz, Antonio H Cruz, Félix Gascón, Isaac Túnez. 2012. "Oxidative stress and inflammation biomarkers in the blood of patients with Huntington's disease." *Neurological research* no. 34 (7):721-724.

[44] Ayala-Peña, Sylvette. 2013. "Role of oxidative DNA damage in mitochondrial dysfunction and Huntington's disease pathogenesis." *Free Radical Biology and Medicine* no. 62:102-110.

[45] Milne, Ginger L, Huiyong Yin, Klarissa D Hardy, Sean S Davies, L Jackson Roberts. 2011. "Isoprostane generation and function." *Chemical reviews* no. 111 (10):5973-5996.

[46] Browne, Susan E, M Flint Beal. 2006. "Oxidative damage in Huntington's disease pathogenesis." *Antioxidants & redox signaling* no. 8 (11-12):2061-2073.

[47] Haun, Florian, Tomohiro Nakamura, Stuart A Lipton. 2013. "Dysfunctional mitochondrial dynamics in the pathophysiology of neurodegenerative diseases." *Journal of cell death* no. 6:27.

[48] Stoy, N, GM Mackay, CM Forrest, J Christofides, M Egerton, TW Stone, LG Darlington. 2005. "Tryptophan metabolism and oxidative stress in patients with Huntington's disease." *Journal of neurochemistry* no. 93 (3):611-623.

[49] Montine, TJ, MF Beal, D Robertson, ME Cudkowicz, I Biaggioni, H O'Donnell, WE Zackert, LJ Roberts, JD Morrow. 1999. "Cerebrospinal fluid F2-isoprostanes are elevated in Huntington's disease." *Neurology* no. 52 (5):1104-1104.

[50] Kralovicova, Silvia, Sarah N Fontaine, Alexandra Alderton, Julia Alderman, K Vala Ragnarsdottir, Steven J Collins, David R Brown. 2009. "The effects of prion protein expression on metal metabolism." *Molecular and Cellular Neuroscience* no. 41 (2):135-147.

[51] Arlt, Sönke, Anatol Kontush, Inga Zerr, Carsten Buhmann, Christian Jacobi, Andreas Schröter, Sigrid Poser, Ulrike Beisiegel. 2002. "Increased lipid peroxidation in cerebrospinal fluid and plasma from patients with Creutzfeldt–Jakob disease." *Neurobiology of disease* no. 10 (2):150-156.

[52] Minghetti, L, F Cardone, A Greco, M Puopolo, G Levi, AJE Green, R Knight, M Pocchiari. 2002. "Increased CSF levels of prostaglandin E2 in variant Creutzfeldt–Jakob disease." *Neurology* no. 58 (1):127-129.

[53] Bleich, S, S Kropp, I Zerr, J Pilz, CH Gleiter, D Degner, M Otto, E Rüther, HA Kretzschmar, J Wiltfang. 2000. "Creutzfeldt–Jakob disease and oxidative stress." *Acta neurologica scandinavica* no. 101 (5):332-334.

[54] Cardenas-Rodriguez, Noemi, Bernardino Huerta-Gertrudis, Liliana Rivera-Espinosa, Hortencia Montesinos-Correa, Cindy Bandala, Liliana Carmona-Aparicio, Elvia Coballase-Urrutia. 2013. "Role of oxidative stress in refractory epilepsy: evidence in patients and experimental models." *International journal of molecular sciences* no. 14 (1):1455-1476.

[55] Hamed, Sherifa A, Moustafa M Abdellah, Nagla El-Melegy. 2004. "Blood levels of trace elements, electrolytes, and oxidative stress/antioxidant systems in epileptic patients." *Journal of pharmacological sciences* no. 96 (4):465-473.

[56] Di Maio, Roberto. 2011. "Neuronal oxidative injury in the development of the epileptic disease: a potential target for novel therapeutic approaches."

[57] Medina, Sonia, Irene De Miguel-Elízaga, Camille Oger, Jean-Marie Galano, Thierry Durand, Miriam Martínez-Villanueva, María Luisa Gil-Del Castillo, Irene Villegas-Martínez, Federico Ferreres, Pedro Martínez-Hernández, Ángel Gil-Izquierdo. 2015. "Dihomo-isoprostanes—nonenzymatic metabolites of AdA—are higher in epileptic patients compared to healthy individuals by a new ultrahigh pressure liquid chromatography–triple quadrupole–tandem mass spectrometry method." *Free Radical Biology and Medicine* no. 79:154-163.

[58] Roberts Ii, L. J., J. P. Fessel, S. S. Davies. 2005. "The biochemistry of the isoprostane, neuroprostane, and isofuran pathways of lipid peroxidation." *Brain Pathology* no. 15 (2):143-148.

[59] Corcoran, T. B., E. Mas, A. E. Barden, T. Durand, J. M. Galano, L. J. Roberts, M. Phillips, K. M. Ho, T. A. Mori. 2011. "Are isofurans and neuroprostanes increased after subarachnoid hemorrhage and traumatic brain injury?" *Antioxidants and Redox Signaling* no. 15 (10):2663-2667. doi: 10.1089/ars.2011.4125.

[60] Hsieh, Y. P., C. L. Lin, A. L. Shiue, H. Yin, J. D. Morrow, J. C. Hsu, T. C. Hsieh, H. J. Wei, H. C. Yen. 2009. "Correlation of F4-neuroprostanes levels in cerebrospinal fluid with outcome of aneurysmal subarachnoid hemorrhage in humans." *Free Radical Biology and Medicine* no. 47 (6):814-824. doi: 10.1016/j. freeradbiomed. 2009.06.026.

[61] Barden, A., L. J. Beilin, J. Ritchie, K. D. Croft, B. N. Walters, C. A. Michael. 1996. "Plasma and urinary 8-iso-prostane as an indicator of lipid peroxidation in pre-eclampsia and normal pregnancy." *Clinical Science* no. 91 (6):711-718.

[62] Barden, A. E., T. B. Corcoran, E. Mas, T. Durand, J. M. Galano, L. J. Roberts, M. Paech, N. A. Muchatuta, M. Phillips, T. A. Mori. 2012. "Is there a role for isofurans and neuroprostanes in pre-eclampsia and normal pregnancy?" *Antioxidants and Redox Signaling* no. 16 (2):165-169. doi: 10.1089/ars.2011.4214.

[63] Bate, Clive, Scott Rutherford, Mike Gravenor, Stuart Reid, Alun Williams. 2002. "Cyclo-oxygenase inhibitors protect against prion-induced neurotoxicity in vitro." *Neuroreport* no. 13 (15):1933-1938.

[64] Pradat, PF, S Attarian, JP Camdessanche, L Carluer, P Cintas, P Corcia, A Echaniz-Laguna, J Gonzalez-Bermejo, N Guy, G Nicolas. 2009. "[Research in amyotrophic lateral sclerosis: what is new in 2009?]." *Revue neurologique* no. 166 (8-9):683-698.

[65] Mitsumoto, Hiroshi, Regina M Santella, Xinhua Liu, Mikhail Bogdanov, Jennifer Zipprich, Hui-Chen Wu, Julie Mahata, Mary Kilty, Kate Bednarz, Daniel Bell. 2008. "Oxidative stress biomarkers in sporadic ALS." *Amyotrophic Lateral Sclerosis* no. 9 (3):177-183.

[66] Consilvio, Christopher, Andrea M Vincent, Eva L Feldman. 2004. "Neuroinflammation, COX-2, and ALS—a dual role?" *Experimental neurology* no. 187 (1):1-10.

[67] Barber, Siân C, Pamela J Shaw. 2010. "Oxidative stress in ALS: key role in motor neuron injury and therapeutic target." *Free Radical Biology and Medicine* no. 48 (5):629-641.

[68] Harris, Sarah G, Josue Padilla, Laura Koumas, Denise Ray, Richard P Phipps. 2002. "Prostaglandins as modulators of immunity." *Trends in immunology* no. 23 (3):144-150.

[69] Wu, Hui-Chen, Qiao Wang, Hwai-I Yang, Habibul Ahsan, Wei-Yann Tsai, Li-Yu Wang, Shu-Yuan Chen, Chien-Jen Chen, Regina M Santella. 2008. "Urinary 15-F2t-isoprostane, aflatoxin B1 exposure and hepatitis B virus infection and hepatocellular carcinoma in Taiwan." *Carcinogenesis* no. 29 (5):971-976.

[70] Rossner, Pavel, Marilie D Gammon, Mary Beth Terry, Meenakshi Agrawal, Fang Fang Zhang, Susan L Teitelbaum, Sybil M Eng, Mia M Gaudet, Alfred I Neugut, Regina M Santella. 2006. "Relationship between urinary 15-F2t-isoprostane and 8-oxodeoxyguanosine levels and breast cancer risk." *Cancer Epidemiology Biomarkers & Prevention* no. 15 (4):639-644.

[71] Epplein, Meira, Adrian A Franke, Robert V Cooney, J Steven Morris, Lynne R Wilkens, Marc T Goodman, Suzanne P Murphy, Brian E Henderson, Laurence N Kolonel, Loïc Le Marchand. 2009. "Association of plasma micronutrient levels and urinary isoprostane with risk of lung cancer: the multiethnic cohort study." *Cancer Epidemiology Biomarkers & Prevention* no. 18 (7):1962-1970.

[72] Brys, Magdalena, Agnieszka Morel, Ewa Forma, Anna Krzeslak, Jacek Wilkosz, Waldemar Rozanski, Beata Olas. 2013. "Relationship of urinary isoprostanes to prostate cancer occurence." *Molecular and cellular biochemistry* no. 372 (1-2):149-153.

[73] Barocas, Daniel A, Saundra Motley, Michael S Cookson, Sam S Chang, David F Penson, Qi Dai, Ginger Milne, L Jackson Roberts, Jason Morrow, Raoul S Concepcion. 2011. "Oxidative stress measured by urine F2-isoprostane level is associated with prostate cancer." *The Journal of urology* no. 185 (6):2102-2107.

[74] Asombang, Akwi W, Violet Kayamba, Mpala Mwanza-Lisulo, Graham Colditz, Victor Mudenda, Kevin Yarasheski, Robert Chott, Deborah C Rubin, C Prakash Gyawali, Edford Sinkala. 2013. "Gastric cancer in Zambian adults: a prospective case-control study that assessed dietary intake and antioxidant status by using urinary isoprostane excretion." *The American Journal of Clinical Nutrition* no. 97 (5):1029-1035.

[75] Siamakpour-Reihani, Sharareh, Peter M Scarbrough, Frances Wang, Ivan Spasojevic, Karel Base, Rebecca Sedjo, Ralph B D'Agostino, Dora Il'yasova. 2012. "Systemic markers of oxidative status and colorectal adenomatous polyps." *Annals of Epidemiology* no. 22 (8):587-591.

[76] Dai, Qi, Yu-Tang Gao, Xiao-Ou Shu, Gong Yang, Ginger Milne, Qiuyin Cai, Wanqing Wen, Nathaniel Rothman, Hui Cai, Honglan Li. 2009. "Oxidative stress, obesity, and breast cancer risk: results from the Shanghai Women's Health Study." *Journal of Clinical Oncology* no. 27 (15):2482-2488.

[77] Thompson, Patricia A, Christine Ambrosone. 2000. "Molecular epidemiology of genetic polymorphisms in estrogen metabolizing enzymes in human breast cancer." *J Natl Cancer Inst Monogr* no. 27:125-134.

[78] Flegal, Katherine M, Margaret D Carroll, Cynthia L Ogden, Lester R Curtin. 2010. "Prevalence and trends in obesity among US adults, 1999-2008." *Jama* no. 303 (3):235-241.

[79] Basu, S. 2010. "Fatty acid oxidation and isoprostanes: Oxidative strain and oxidative stress." *Prostaglandins Leukotrienes and Essential Fatty Acids* no. 82 (4-6):219-225. doi: 10.1016/j.plefa.2010.02.031.

[80] Morrow, Jason D. 2005. "Quantification of isoprostanes as indices of oxidant stress and the risk of atherosclerosis in humans." *Arteriosclerosis, thrombosis, and vascular biology* no. 25 (2):279-286.

[81] Vincent, Heather K, Ann G Taylor. 2006. "Biomarkers and potential mechanisms of obesity-induced oxidant stress in humans." *International journal of obesity* no. 30 (3):400-418.

[82] Vincent, Heather K, Kim E Innes, Kevin R Vincent. 2007. "Oxidative stress and potential interventions to reduce oxidative stress in

overweight and obesity." *Diabetes, Obesity and Metabolism* no. 9 (6):813-839.

[83] Dow, Caitlin A, Betsy C Wertheim, Bhimanagouda S Patil, Cynthia A Thomson. 2013. "Daily consumption of grapefruit for 6 weeks reduces urine F2-isoprostanes in overweight adults with high baseline values but has no effect on plasma high-sensitivity C-reactive protein or soluble vascular cellular adhesion molecule 1." *The Journal of nutrition* no. 143 (10):1586-1592.

[84] Keaney, John F, Martin G Larson, Ramachandran S Vasan, Peter WF Wilson, Izabella Lipinska, Diane Corey, Joseph M Massaro, Patrice Sutherland, Joseph A Vita, Emelia J Benjamin. 2003. "Obesity and systemic oxidative stress clinical correlates of oxidative stress in the Framingham Study." *Arteriosclerosis, thrombosis, and vascular biology* no. 23 (3):434-439.

[85] Kaikkonen, Jari E, Teemu Vilppo, Janne Asikainen, Sari Voutilainen, Sudhir Kurl, Jukka T Salonen. 2013. "Fatty acids as determinants of in-vivo lipid peroxidation: The EFFGE study in Eastern Finnish hypertensive and non-hypertensive subjects." *Annals of medicine* no. 45 (5-6):455-464.

[86] Crist, Matthew B, Vlada V Melekhin, Aihua Bian, Ayumi Shintani, Ginger L Milne, Asha R Kallianpur, Leigh Anne Dageforde, David W Haas, Todd Hulgan. 2013. "Higher serum iron is associated with increased oxidant stress in HIV-infected men." *Journal of acquired immune deficiency syndromes (1999)* no. 64 (4).

[87] Dennis, B Adam, Adviye Ergul, Barbara A Gower, Jerry D Allison, Catherine L Davis. 2013. "Oxidative stress and cardiovascular risk in overweight children in an exercise intervention program." *Childhood Obesity* no. 9 (1):15-21.

[88] Alkazemi, Dalal, Grace M Egeland, Lyman Jackson Roberts, Stan Kubow. 2012. "Isoprostanes and isofurans as non-traditional risk factors for cardiovascular disease among Canadian Inuit." *Free Radical Research* no. 46 (10):1258-1266.

[89] Il'yasova, Dora, Frances Wang, Ivan Spasojevic, Karel Base, Ralph B D'agostino, Lynne E Wagenknecht. 2012. "Racial Differences in Urinary F2-Isoprostane Levels and the Cross-Sectional Association With bmi." *Obesity* no. 20 (10):2147-2150.

[90] Billings, Frederic T, Mias Pretorius, Jonathan S Schildcrout, Nathaniel D Mercaldo, John G Byrne, T Alp Ikizler, Nancy J Brown. 2012.

"Obesity and oxidative stress predict AKI after cardiac surgery." *Journal of the American Society of Nephrology* no. 23 (7):1221-1228.

[91] Buchowski, Maciej S, Nobuko Hongu, Sari Acra, Li Wang, Joshua Warolin, L Jackson Roberts. 2012. "Effect of modest caloric restriction on oxidative stress in women, a randomized trial." *PloS one* no. 7 (10):e47079.

[92] Dorjgochoo, Tsogzolmaa, Yu-Tang Gao, Wong-Ho Chow, Xiao-ou Shu, Gong Yang, Quiyin Cai, Nathaniel Rothman, Hui Cai, Honglan Li, Xinqing Deng. 2011. "Obesity, age, and oxidative stress in middle-aged and older women." *Antioxidants & redox signaling* no. 14 (12):2453-2460.

[93] Kanaya, Alka M, Christina L Wassel, Pamela J Stoddard, Tamara B Harris, Steven R Cummings, Stephen B Kritchevsky, Bret H Goodpaster, Christie Green, Suzanne Satterfield, Myron D Gross. 2011. "F2-Isoprostanes and Adiposity in Older Adults." *Obesity* no. 19 (4):861-867.

[94] Pfeuffer, Maria, Kerstin Fielitz, Christiane Laue, Petra Winkler, Diana Rubin, Ulf Helwig, Katrin Giller, Julia Kammann, Edzard Schwedhelm, Rainer H Böger. 2011. "CLA does not impair endothelial function and decreases body weight as compared with safflower oil in overweight and obese male subjects." *Journal of the American College of Nutrition* no. 30 (1):19-28.

[95] Araki, Shunsuke, Kazushige Dobashi, Yukiyo Yamamoto, Kohtaro Asayama, Koichi Kusuhara. 2010. "Increased plasma isoprostane is associated with visceral fat, high molecular weight adiponectin, and metabolic complications in obese children." *European journal of pediatrics* no. 169 (8):965-970.

[96] Tadros, Thomas M, Joseph M Massaro, Guido A Rosito, Udo Hoffmann, Ramachandran S Vasan, Martin G Larson, John F Keaney, Izabella Lipinska, James B Meigs, Sekar Kathiresan. 2010. "Pericardial fat volume correlates with inflammatory markers: the Framingham Heart Study." *Obesity* no. 18 (5):1039-1045.

[97] Crist, Betsy L, D Lee Alekel, Laura M Ritland, Laura N Hanson, Ulrike Genschel, Manju B Reddy. 2009. "Association of oxidative stress, iron, and centralized fat mass in healthy postmenopausal women." *Journal of Women's Health* no. 18 (6):795-801.

[98] Redhage, Leigh Anne, Ayumi Shintani, David W Haas, Nkiruka Emeagwali, Milica Markovic, Ikwo Oboho, Christopher Mwenya, Husamettin Erdem, Edward P Acosta, Jason D Morrow. 2009. "Clinical

factors associated with plasma F2-isoprostane levels in HIV-infected adults." *HIV clinical trials* no. 10 (3):181-192.

[99] Pou, Karla M, Joseph M Massaro, Udo Hoffmann, Ramachandran S Vasan, Pal Maurovich-Horvat, Martin G Larson, John F Keaney, James B Meigs, Izabella Lipinska, Sekar Kathiresan. 2007. "Visceral and subcutaneous adipose tissue volumes are cross-sectionally related to markers of inflammation and oxidative stress the Framingham heart study." *Circulation* no. 116 (11):1234-1241.

[100] Komakula, Sushma, Sumita Khatri, Joel Mermis, Samira Savill, Shireen Haque, Mauricio Rojas, LouAnn Brown, Gerald W Teague, Fernando Holguin. 2007. "Body mass index is associated with reduced exhaled nitric oxide and higher exhaled 8-isoprostanes in asthmatics." *Respiratory research* no. 8 (1):32.

[101] Urakawa, Hideki, Akira Katsuki, Yasuhiro Sumida, Esteban C Gabazza, Shuichi Murashima, Kohei Morioka, Noriko Maruyama, Nagako Kitagawa, Takashi Tanaka, Yasuko Hori. 2003. "Oxidative stress is associated with adiposity and insulin resistance in men." *The Journal of Clinical Endocrinology & Metabolism* no. 88 (10):4673-4676.

[102] Warolin, Joshua, Kimberly R Coenen, Jeff L Kantor, Lauren E Whitaker, Li Wang, Sari A Acra, L Jackson Roberts, MS Buchowski. 2014. "The relationship of oxidative stress, adiposity and metabolic risk factors in healthy Black and White American youth." *Pediatric obesity* no. 9 (1):43-52.

[103] Dorjgochoo, Tsogzolmaa, Yu-Tang Gao, Wong-Ho Chow, Xiao-ou Shu, Gong Yang, Qiuyin Cai, Nathaniel Rothman, Hui Cai, Honglan Li, Xinqing Deng. 2012. "Major metabolite of F2-isoprostane in urine may be a more sensitive biomarker of oxidative stress than isoprostane itself." *The American Journal of Clinical Nutrition* no. 96 (2):405-414.

[104] Davì, Giovanni, Angela Falco, Carlo Patrono. 2004. "Determinants of F 2-isoprostane biosynthesis and inhibition in man." *Chemistry and physics of lipids* no. 128 (1):149-163.

[105] Roberts, L Jackson, John A Oates, MacRae F Linton, Sergio Fazio, Beth P Meador, Myron D Gross, Yu Shyr, Jason D Morrow. 2007. "The relationship between dose of vitamin E and suppression of oxidative stress in humans." *Free Radical Biology and Medicine* no. 43 (10):1388-1393.

[106] Block, Gladys, Christopher D Jensen, Jason D Morrow, Nina Holland, Edward P Norkus, Ginger L Milne, Mark Hudes, Tapashi B Dalvi, Patricia B Crawford, Ellen B Fung. 2008. "The effect of vitamins C and

E on biomarkers of oxidative stress depends on baseline level." *Free Radical Biology and Medicine* no. 45 (4):377-384.

[107] Hoskins, Aimee, Jackson L Roberts, Ginger Milne, Leena Choi, Ryszard Dworski. 2012. "Natural-source d-α-tocopheryl acetate inhibits oxidant stress and modulates atopic asthma in humans in vivo." *Allergy* no. 67 (5):676-682.

[108] Espey, Michael Graham, Ping Chen, Brian Chalmers, Jeanne Drisko, Andrew Y Sun, Mark Levine, Qi Chen. 2011. "Pharmacologic ascorbate synergizes with gemcitabine in preclinical models of pancreatic cancer." *Free Radical Biology and Medicine* no. 50 (11):1610-1619.

[109] Traustadóttir, Tinna, Sean S Davies, Anthoney A Stock, Yali Su, Christopher B Heward, L Jackson Roberts, S Mitchell Harman. 2009. "Tart cherry juice decreases oxidative stress in healthy older men and women." *The Journal of nutrition* no. 139 (10):1896-1900.

[110] McAnulty, Lisa S, David C Nieman, Charles L Dumke, Lesli A Shooter, Dru A Henson, Alan C Utter, Ginger Milne, Steven R McAnulty. 2011. "Effect of blueberry ingestion on natural killer cell counts, oxidative stress, and inflammation prior to and after 2.5 h of running." *Applied Physiology, Nutrition, and Metabolism* no. 36 (6):976-984.

[111] Khan, Faisel, Sumantra Ray, Angela M Craigie, Gwen Kennedy, Alexander Hill, Karen L Barton, Jane Broughton, Jill JF Belch. 2014. "Lowering of oxidative stress improves endothelial function in healthy subjects with habitually low intake of fruit and vegetables: A randomized controlled trial of antioxidant-and polyphenol-rich blackcurrant juice." *Free Radical Biology and Medicine* no. 72:232-237.

[112] Meyer, Katie A, Femke PC Sijtsma, Jennifer A Nettleton, Lyn M Steffen, Linda Van Horn, James M Shikany, Myron D Gross, Jaakko Mursu, Maret G Traber, David R Jacobs. 2013. "Dietary patterns are associated with plasma F 2-isoprostanes in an observational cohort study of adults." *Free Radical Biology and Medicine* no. 57:201-209.

[113] Rink, Stephanie M, Pauline Mendola, Sunni L Mumford, Jill K Poudrier, Richard W Browne, Jean Wactawski-Wende, Neil J Perkins, Enrique F Schisterman. 2013. "Self-report of fruit and vegetable intake that meets the 5 a day recommendation is associated with reduced levels of oxidative stress biomarkers and increased levels of antioxidant defense in premenopausal women." *Journal of the Academy of Nutrition and Dietetics* no. 113 (6):776-785.

[114] Jiang, Yu, Sheng-Hui Wu, Xiao-Ou Shu, Yong-Bing Xiang, Bu-Tian Ji, Ginger L Milne, Qiuyin Cai, Xianglan Zhang, Yu-Tang Gao, Wei Zheng. 2014. "Cruciferous vegetable intake is inversely correlated with circulating levels of proinflammatory markers in women." *Journal of the Academy of Nutrition and Dietetics* no. 114 (5):700-708. e2.

[115] Barden, Anne E, Trevor A Mori, Janet A Dunstan, Angela L Taylor, Catherine A Thornton, Kevin D Croft, Lawrence J Beilin, Susan L Prescott. 2004. "Fish oil supplementation in pregnancy lowers F2-isoprostanes in neonates at high risk of atopy." *Free Radical Research* no. 38 (3):233-239.

[116] Kiecolt-Glaser, Janice K, Elissa S Epel, Martha A Belury, Rebecca Andridge, Jue Lin, Ronald Glaser, William B Malarkey, Beom Seuk Hwang, Elizabeth Blackburn. 2013. "Omega-3 fatty acids, oxidative stress, and leukocyte telomere length: a randomized controlled trial." *Brain, behavior, and immunity* no. 28:16-24.

[117] Signorini, Cinzia, Claudio De Felice, Silvia Leoncini, Thierry Durand, Jean-Marie Galano, Alessio Cortelazzo, Gloria Zollo, Roberto Guerranti, Stefano Gonnelli, Carla Caffarelli. 2014. "Altered erythrocyte membrane fatty acid profile in typical Rett syndrome: Effects of omega-3 polyunsaturated fatty acid supplementation." *Prostaglandins, Leukotrienes and Essential Fatty Acids (PLEFA)* no. 91 (5):183-193.

[118] De Felice, C., C. Signorini, T. Durand, C. Oger, A. Guy, V. Bultel-Poncé, J. M. Galano, L. Ciccoli, S. Leoncini, M. D'Esposito, S. Filosa, A. Pecorelli, G. Valacchi, J. Hayek. 2011. "F2-dihomo-isoprostanes as potential early biomarkers of lipid oxidative damage in Rett syndrome." *Journal of lipid research* no. 52 (12):2287-2297. doi: 10.1194/jlr.P017798.

[119] Signorini, C., C. De Felice, S. Leoncini, A. Giardini, M. D'Esposito, S. Filosa, F. Della Ragione, M. Rossi, A. Pecorelli, G. Valacchi, L. Ciccoli, J. Hayek. 2011. "F4-neuroprostanes mediate neurological severity in Rett syndrome." *Clinica Chimica Acta* no. 412 (15-16):1399-1406. doi: 10.1016/j.cca.2011.04.016.

[120] Gorrindo, Phillip, Christianne Joy Lane, Evon Batey Lee, B McLaughlin, Pat Levitt. 2013. "Enrichment of elevated plasma F2t-isoprostane levels in individuals with autism who are stratified by presence of gastrointestinal dysfunction." *PloS one* no. 8 (7):e68444.

[121] Ming, X, TP Stein, M Brimacombe, WG Johnson, GH Lambert, GC Wagner. 2005. "Increased excretion of a lipid peroxidation biomarker in

autism." *Prostaglandins, leukotrienes and essential fatty acids* no. 73 (5):379-384.

[122] Korade, Zeljka, Libin Xu, Karoly Mirnics, Ned A Porter. 2013. "Lipid biomarkers of oxidative stress in a genetic mouse model of Smith-Lemli-Opitz syndrome." *Journal of inherited metabolic disease* no. 36 (1):113-122.

[123] Akohoue, Sylvie A, Sadhna Shankar, Ginger L Milne, Jason Morrow, Kong Y Chen, Wilfred U Ajayi, Maciej S Buchowski. 2007. "Energy expenditure, inflammation, and oxidative stress in steady-state adolescents with sickle cell anemia." *Pediatric Research* no. 61 (2):233-238.

[124] Collins, Clare E, Paul Quaggiotto, Lisa Wood, Edward V O'loughlin, Richard L Henty, Manohar L Garg. 1999. "Elevated plasma levels of F2α isoprostane in cystic fibrosis." *Lipids* no. 34 (6):551-556.

[125] Montuschi, Paolo, Sergei A Kharitonov, Giovanni Ciabattoni, Massimo Corradi, Liesbeth van Rensen, Duncan M Geddes, Margaret E Hodson, Peter J Barnes. 2000. "Exhaled 8-isoprostane as a new non-invasive biomarker of oxidative stress in cystic fibrosis." *Thorax* no. 55 (3):205-209.

[126] Perrone, S, M Longini, CV Bellieni, G Centini, A Kenanidis, L De Marco, F Petraglia, G Buonocore. 2007. "Early oxidative stress in amniotic fluid of pregnancies with Down syndrome." *Clinical Biochemistry* no. 40 (3):177-180.

[127] De Felice, Claudio, Lucia Ciccoli, Silvia Leoncini, Cinzia Signorini, Marcello Rossi, Laura Vannuccini, Gianni Guazzi, Giuseppe Latini, Mario Comporti, Giuseppe Valacchi. 2009. "Systemic oxidative stress in classic Rett syndrome." *Free Radical Biology and Medicine* no. 47 (4):440-448.

[128] Galano, Jean-Marie, Emilie Mas, Anne Barden, Trevor A Mori, Cinzia Signorini, Claudio De Felice, Aaron Barrett, Catherine Opere, Edith Pinot, Edzard Schwedhelm. 2013. "Isoprostanes and neuroprostanes: total synthesis, biological activity and biomarkers of oxidative stress in humans." *Prostaglandins & other lipid mediators* no. 107:95-102.

[129] Cracowski, Jean-Luc, Thierry Durand. 2006. "Cardiovascular pharmacology and physiology of the isoprostanes." *Fundamental & Clinical Pharmacology* no. 20 (5):417-427. doi: 10.1111/j.1472-8206.2006.00435.x.

[130] Ting, Harold J, Fadi T Khasawneh. 2010. "Platelet function and Isoprostane biology. Should Isoprostanes be the newest member of the Orphan-ligand family?" *Journal of Biomedical Science* no. 17 (1):24.
[131] Kumar, A., E. Kingdon, J. Norman. 2005. "The isoprostane 8-iso-PGF2α suppresses monocyte adhesion to human microvascular endothelial cells via two independent mechanisms." *FASEB Journal* no. 19 (3):443-445. doi: 10.1096/fj.03-1364fje.
[132] Yura, Takafumi, Megumu Fukunaga, Rizwan Khan, George N. Nassar, Kamal F. Badr, Angel Montero. 1999. "Free-radical–generated F2-isoprostane stimulates cell proliferation and endothelin-1 expression on endothelial cells." *Kidney International* no. 56 (2):471-478. doi: http://dx.doi.org/10.1046/j.1523-1755.1999.00596.x.
[133] Miggin, Sinead M., B. Therese Kinsella. 2001. "Thromboxane A2 receptor mediated activation of the mitogen activated protein kinase cascades in human uterine smooth muscle cells." *Biochimica et Biophysica Acta (BBA) - Molecular Cell Research* no. 1539 (1–2):147-162. doi: http://dx.doi.org/10.1016/S0167-4889(01)00103-3.
[134] Bauer, J., A. Ripperger, S. Frantz, S. Ergün, E. Schwedhelm, R. A. Benndorf. 2014. "Pathophysiology of isoprostanes in the cardiovascular system: Implications of isoprostane-mediated thromboxane A2 receptor activation." *British Journal of Pharmacology* no. 171 (13):3115-3131.
[135] Zhao, Min, ChristopherJ Destache, SunnyE Ohia, CatherineA Opere. 2009. "Role of Prostanoid Production and Receptors in the Regulation of Retinal Endogenous Amino Acid Neurotransmitters by 8-Isoprostaglandin E2, Ex Vivo." *Neurochemical Research* no. 34 (12):2170-2180. doi: 10.1007/s11064-009-0013-x.
[136] Joy, Andrew P., Elizabeth A. Cowley. 2008. "8-iso-PGE2 Stimulates Anion Efflux from Airway Epithelial Cells via the EP4 Prostanoid Receptor." *American Journal of Respiratory Cell and Molecular Biology* no. 38 (2):143-152. doi: 10.1165/rcmb.2006-0295OC.
[137] Seto, Victoria, Christina Hirota, Simon Hirota, Luke J. Janssen. 2008. "E-Ring Isoprostanes Stimulate a Cl− Conductance in Airway Epithelium via Prostaglandin E2-Selective Prostanoid Receptors." *American Journal of Respiratory Cell and Molecular Biology* no. 38 (1):88-94. doi: 10.1165/rcmb.2007-0117OC.
[138] Clarke, D. L., M. G. Belvisi, E. Hardaker, R. Newton, M. A. Giembycz. 2005. "E-ring 8-isoprostanes are agonists at EP 2- and EP 4- prostanoid receptors on human airway smooth muscle cells and regulate the release

of colony-stimulating factors by activating cAMP-dependent protein kinase." *Molecular Pharmacology* no. 67 (2):383-393.

[139] Paredes, Christine, Tracy Tazzeo, Luke J. Janssen. 2007. "E-Ring Isoprostane Augments Cholinergic Neurotransmission in Bovine Trachealis via FP Prostanoid Receptors." *American Journal of Respiratory Cell and Molecular Biology* no. 37 (6):739-747. doi: 10.1165/rcmb.2007-0022OC.

[140] Weber, T. J., L. M. Markillie. 2003. "Regulation of activator protein-1 by 8-iso-prostaglandin E2 in a thromboxane A2 receptor-dependent and -independent manner." *Molecular Pharmacology* no. 63 (5):1075-1081.

[141] Clarke, Deborah L., Mark A. Giembycz, Hema J. Patel, Maria G. Belvisi. 2004. "E-Ring 8-isoprostanes inhibit ACh release from parasympathetic nerves innervating guinea-pig trachea through agonism of prostanoid receptors of the EP3-subtype." *British Journal of Pharmacology* no. 141 (4):600-609. doi: 10.1038/sj.bjp.0705648.

[142] Janssen, L. J., T. Tazzeo. 2002. "Involvement of TP and EP3 receptors in vasoconstrictor responses to isoprostanes in pulmonary vasculature." *Journal of Pharmacology and Experimental Therapeutics* no. 301 (3):1060-1066. doi: 10.1124/jpet.301.3.1060.

[143] Sametz, Wolfgang, Simone Hennerbichler, Sonja Glaser, Reinhold Wintersteiger, Heinz Juan. 2000. "Characterization of prostanoid receptors mediating actions of the isoprostanes, 8-iso-PGE2 and 8-iso-PGF2α, in some isolated smooth muscle preparations." *British Journal of Pharmacology* no. 130 (8):1903-1910. doi: 10.1038/sj.bjp.0703522.

[144] Elmhurst, J. L., P. A. Betti, P. K. Rangachari. 1997. "Intestinal effects of isoprostanes: Evidence for the involvement of prostanoid EP and TP receptors." *Journal of Pharmacology and Experimental Therapeutics* no. 282 (3):1198-1205.

[145] Van Der Sterren, S., L. Kessels, F. Perez-Vizcaino, A. L. Cogolludo, E. Villamor. 2014. "Prenatal exposure to hyperoxiamodifies the thromboxane prostanoid receptor-mediated response to H2O2 in the ductus arteriosus of the chicken embryo." *Journal of Physiology and Pharmacology* no. 65 (2):283-293.

[146] van der Sterren, Saskia, Eduardo Villamor. 2011. "Contractile effects of 15-E2t-isoprostane and 15-F2t-isoprostane on chicken embryo ductus arteriosus." *Comparative Biochemistry and Physiology Part A: Molecular & Integrative Physiology* no. 159 (4):436-444. doi: http://dx.doi.org/10.1016/j.cbpa.2011.04.010.

[147] Evans, Angela R, Heidi Junger, Michael D Southall, Grant D Nicol, Linda S Sorkin, James T Broome, Timothy W Bailey, Michael R Vasko. 2000. "Isoprostanes, novel eicosanoids that produce nociception and sensitize rat sensory neurons." *Journal of Pharmacology and Experimental Therapeutics* no. 293 (3):912-920.

[148] Comporti, Mario, Cinzia Signorini, Silvia Leoncini, Giuseppe Buonocore, Viviana Rossi, Lucia Ciccoli. 2004. "Plasma F2-isoprostanes are elevated in newborns and inversely correlated to gestational age." *Free Radical Biology and Medicine* no. 37 (5):724-732. doi: http://dx.doi.org/10.1016/j.freeradbiomed.2004.06.007.

[149] Cracowski, Jean-Luc, Françoise Stanke-Labesque, Philippe Devillier, Olivier Chavanon, Mark Hunt, Claude Souvignet, Germain Bessard. 2000. "Human internal mammary artery contraction by isoprostaglandin F2α type-III (8-iso-prostaglandin F2α)." *European Journal of Pharmacology* no. 397 (1):161-168. doi: http://dx.doi.org/ 10.1016/S0014-2999(00)00217-X.

[150] Marlière, Stéphanie, Jean-Luc Cracowski, Thierry Durand, Olivier Chavanon, Janine Bessard, Alexandre Guy, Françoise Stanke-Labesque, Jean-Claude Rossi, Germain Bessard. 2002. "The 5-series F2-isoprostanes possess no vasomotor effects in the rat thoracic aorta, the human internal mammary artery and the human saphenous vein." *British Journal of Pharmacology* no. 135 (5):1276-1280. doi: 10.1038/sj.bjp.0704558.

[151] Greaves, K., S. R. Dixon, I. O. Coker, A. I. Mallet, M. Avkiran, M. J. Shattock, M. J. Fejka, W. W. O'Neill, R. Senior, S. Redwood, M. S. Marber. 2004. "Influence of isoprostane F2α-III on reflow after myocardial infarction." *European Heart Journal* no. 25 (10):847-853. doi: 10.1016/j.ehj.2004.03.015.

[152] Tanito, Masaki, Richard S Brush, Michael H Elliott, Lea D Wicker, Kimberly R Henry, Robert E Anderson. 2009. "High levels of retinal membrane docosahexaenoic acid increase susceptibility to stress-induced degeneration." *Journal of lipid research* no. 50 (5):807-819.

[153] Opere, Catherine A, Kim Ford, Min Zhao, Sunny E Ohia. 2008. "Regulation of neurotransmitter release from ocular tissues by isoprostanes." *Methods and findings in experimental and clinical pharmacology* no. 30 (9):697-701.

[154] Izzotti, Alberto, Alessandro Bagnis, Sergio C. Saccà. 2006. "The role of oxidative stress in glaucoma." *Mutation Research/Reviews in Mutation*

Research no. 612 (2):105-114. doi: http://dx.doi.org/10.1016/j.mrrev.2005.11.001.

[155] Jamil, Jamal, Ashley Wright, Edem Kegey, Arnecia Faye Flowers, Namonique Jarell Flyod, Casey Kotera, Alexandre Guy, Jean-Marie Galano, Thierry Durand, Ya Fatou Njie-Mbye. 2012. "Regulation of [3H] D-aspartate release by the 5-F2t-isoprostane and its 5-epimer in isolated bovine retina." *Neurochemical research* no. 37 (3):574-582.

[156] Mozaffarian, Dariush, Jason H. Y. Wu. 2011. "Omega-3 Fatty Acids and Cardiovascular Disease: Effects on Risk Factors, Molecular Pathways, and Clinical Events." *Journal of the American College of Cardiology* no. 58 (20):2047-2067. doi: http://dx.doi.org/10.1016/j.jacc.2011.06.063.

[157] Mozaffarian, Dariush, Jason HY Wu. 2011. "Omega-3 fatty acids and cardiovascular disease: effects on risk factors, molecular pathways, and clinical events." *Journal of the American College of Cardiology* no. 58 (20):2047-2067.

[158] Saravanan, Palaniappan, Neil C. Davidson, Erik B. Schmidt, Philip C. Calder. "Cardiovascular effects of marine omega-3 fatty acids." *The Lancet* no. 376 (9740):540-550. doi: http://dx.doi.org/10.1016/S0140-6736(10)60445-X.

[159] Jude, S, S Bedut, S Roger, M Pinault, P Champeroux, E White, J-Y Le Guennec. 2003. "Peroxidation of docosahexaenoic acid is responsible for its effects on ITO and ISS in rat ventricular myocytes." *British Journal of Pharmacology* no. 139 (4):816-822.

ABOUT THE AUTHORS

Dr. Javier Marhuenda Hernández

Faculty of Health Sciences

Food Technology and Nutrition Department

Catholic University of San Antonio, Murcia, Spain

Tel: 968278618; Email: jmarhuenda@ucam.edu

Dr. Javier Marhuenda Hernández is currently a research scientist of the group "Nutrition, oxidative stress and bioavailability" at the Catholic University of San Antonio (UCAM). He has developed his career in CEBAS-CSIC and UCAM. He has researched about the antioxidant capacity of different food matrices, highlighting wine and grape juice. The majority of his career has been dedicated to the determination of oxidative stress biomarkers of lipid peroxidation, besides other biomarkers as DNA oxidative injury metabolites.

Dra. Débora Villaño Valencia, PhD

Bachelor in Pharmacy

Faculty of Health Sciences

Catholic University of San Antonio. Murcia, Spain

Tel: 968278618; Email: dvillano@ucam.edu

Dra. Débora Villaño Valencia has a degree in Pharmacy from the University of Navarra and a PhD in Pharmacy from the University of Seville in the Area of Nutrition and Food Science. She has been a post-doctoral researcher at the IRCCS San Raffaele INRAN and Rome Institutions. She has published numerous articles and book chapters SCI related to the effect of plant compounds in diseases with oxidative and inflammatory component. She has extensive experience in managing research results, patents and technology transfer in the food sector. Dr. Villaño Valencia currently works as professor in the Degree of Pharmacy at the Catholic University of San Antonio (UCAM) and conducts her research in the Group Nutrition Group, Oxidative Stress and Bioavailability of UCAM. Her lines of work include the characterization of phytochemicals (mainly polyphenols) present in plant foods and evaluation of their biological effects in studies of nutritional intervention.

Dra. Begoña Cerdá, PhD
Bachelor in Pharmacy
Faculty of Health Sciences
Catholic University of San Antonio, Murcia, Spain
Tel: 968278618; Email: bcerda@ucam.edu

Dra. Begoña Cerdá is a research scientist at the Catholic University of San Antonio (UCAM). Her career was developed in CEBAS-CSIC, Universidad Europea de Madrid and UCAM. She researched about bioavailability and antioxidant capacity of different vegetable product like pomegranate, walnuts, strawberrys or wine. The main of the research has been determination of oxidative stress biomarkers after ingestion of vegetable products in different organic fluids.

Dra. Pilar Zafrilla Rentero, PhD
Bachelor in Pharmacy
Faculty of Health Sciences
Catholic University of San Antonio, Murcia, Spain
Tel: 968278618; Mail: mpzafrilla@ucam.edu

Dra. Pilar Zafrilla Rentero is a principal researcher of the group "Nutrition, oxidative stress and bioavailability" at the Catholic University of San Antonio. The groups' focus is on the research of oxidative stress and sport, as well as oxidative stress in Alzheimer's disease at different stages, analyzing different biomarkers, antioxidant enzymes, inflammatory markers and total antioxidant capacity. They have continued their research on this sense, incorporating diseases such as metabolic syndrome, fragile and macular degeneration. Another currently research line is marked by our collaboration with the research group of quality, safety and bioactivity of plant foods from CEBAS-CSIC, conducting clinical trials and determining the bioavailability of different bioactive substances. Moreover, they analyze the effect of these substances on the expression of various biomarkers of oxidative stress, inflammation and lipid profile, thanks to contracts with various national and international companies.

INDEX

A

B

C

D

E

F

G

H

I

K

L

M

N

O

P

R

S